Mobile Prussia

Ottmar Ette

Mobile Prussia

Views Beyond the National

palgrave
macmillan

Ottmar Ette
Potsdam, Germany

This book is a translation of the original German edition, "Mobile Preußen" by Ette, Ottmar, published by Springer-Verlag GmbH, DE in 2019. The translation was done with the help of artificial intelligence (machine translation by the service DeepL.com). A subsequent human revision was done primarily in terms of content, so that the book will read stylistically differently from a conventional translation. Springer Nature works continuously to further the development of tools for the production of books and on the related technologies to support the authors.

ISBN 978-3-476-05838-6 ISBN 978-3-476-05839-3 (eBook)
https://doi.org/10.1007/978-3-476-05839-3

This Palgrave Macmillan imprint is published by the registered company Springer-Verlag GmbH, DE part of Springer Nature.
The registered company address is: Heidelberger Platz 3, 14197 Berlin, Germany

In the vast park of Sanssouci
the most beautiful place and its
silent inhabitant dedicated.

Preface: Views of Prussia, Beyond the National

In principle, Prussia can be approached conceptually and constructively in at least three different ways. Firstly, there is the territorial dimension: one can understand Prussia as a specific territory and thus as a territorial state with its area, its borders, its population, indeed its history. We are then dealing with Prussia as a nation state, but also as a cultural state or certainly as a military power, which can then be analysed further. This is undoubtedly the most common approach: It generally traces the history and histories of this territory, examines, for example, its social, cultural or economic aspects, and assumes a more or less high degree of homogeneity within it. Prussia can thus be plotted on maps as a territorial state and endowed with very specific characteristics that give this state its identity.

Secondly, Prussia could be examined in its relational dimension: Prussia then appears embedded in a relational network of relations with other states and nations, functions with regard to different politics, economic or social relations, and with regard to its cultures and languages not as an essence in the sense of a 'true' Prussia and its identity, but as a relation that can be functionally determined. Under these auspices, one can, for example, indicate foreign trade or foreign policy, the relative importance of cultural or scientific institutions and ask about their international networking, report on alliance policies or warlike conflicts that shape the essence of this body of state and its institutions. Again, this could be mapped: Prussia would also be *on the map in* this sense.

Thirdly, however, beyond spatial-historical models, one can identify a mobile dimension of the history of movement: it includes the territorial as well as the relational, but in both it places the processual, the unfinished and unsettled at the center of interest and understands a space starting from all those movements that cross it and thus constitute it in the true sense by means of these crossings. This is precisely what is meant by the dimension of the history of movement, which captures, for example, a Prussia in movement and out of movement. It is not about a being and an essence, not about an identity and its variants, but about the processual and dynamic in all phenomena. But how would such a history of movement be redeemed?

Prussia would then appear less as a territorial state with its fixed borders and boundaries, with its (asserted) identity and its (assumed) homogeneity, but rather as an entity in a state of constant reconstruction, whose developmental processes are as open as its ongoing transformations, constantly crossed by all those transfers that turn a given territory with its network of relationships into a mobile, which is in a state of irreversible movement in political as well as economic, in social as well as cultural, in linguistic as well as artistic respects.

Identity would then only be that enforced holding of a certain situation, that reduction of all thinkable and liveable affiliations, which are brought to a territory, a country or even a cultural nation from the side of interest, in order to stop all movements. One could perhaps imagine this most vividly as a historical daguerreotype with a very long exposure time, which would not allow any movement, if the resulting image is not to be 'blurred'. In the photographer's mind, with its central perspective, this image already existed in advance.

From such a still image, the path to the deadly identities would not be (and indeed was historically)[1] far, which undertake a radical reduction of all manifold affiliations in such a way that only one affiliation—even if it were of a fictitious kind—dominates everything else. The conceptual and ultimately normative guidelines for this daguerreotype would be decisive, but not the life of the object itself, which would be *fixed and* strangely frozen in time and space.

If we decide in favor of the third dimension—and this is the path this book follows—the result is a Prussia of many cultures and languages, a Prussia of many migrations and inclusions, a Prussia of many origins as well as of many futures, many of which—and also highly important ones—have not been realized historically and also never seem to be realized anymore. There is no essence of the Prussian here, not even a fixed constellation of what we want to call Prussia, even in a relational sense.

Which image would correspond to the spatial-historical daguerreotype on the level of the history of movement? Well, we would have and have to do with a mobile and constantly dynamic configuration in which the figures move like the parts of a mobile and stand in constantly new relations to one another. In addition, there would be an observer's perspective that renounces any central fixation and makes multi-perspective recordings in which the phenomena are portrayed from different angles at the same time. In this way, however, a Prussia would be contoured that is made of movements, that consciously inscribes the fourth dimension of time into the three dimensions of space and from this allows a multi-dimensional design to emerge that can no longer simply be fixed and still. A thinking space would emerge from the in-between of all these movements, which would certainly not bring back the Prussia that had become historical, but would keep Prussia in its historical becoming ready and available for a future that is still open.

In these *views*—and in the subtitle Georg Forster's *Ansichten vom Niederrhein* as well as Alexander von Humboldt's *Ansichten der Natur* or Adelbert von Chamisso's views went into multiple perspectives—*jenseits des Nationalen* it is precisely about such a view of Prussia that does not focus on the nation state. Rather, it is about the greatest possible, indeed radical, openness of all those *figures* and personalities who populate the pages of this book and, in an exemplary manner, stake out the possible futures that characterized Prussia during the eighteenth century and well into the nineteenth century.

[1] Cf. Maalouf, Amin: *Les Identités meurtrières*. Paris: Editions Grasset & Fasquelle 2005.

In other words: How would a territory have developed in which an Anton Wilhelm Amo could have made himself at home and become part of the Prussian university body? What opportunities would a society have offered in which Heinrich von Kleist's lovers could have found and united beyond all ethnic and religious boundaries, in a community in which Kleist would not have sought suicide in the manner of his characters? How could that nation-state have been formed which a Rahel Levin Varnhagen dreamed of in her attic truths before and after Napoleon's invasion of Berlin and the temporary downfall of Prussia? What kind of conviviality would have come about if an Adelbert von Chamisso had been able to bring in all those aspects he had collected on his journey around the world in the seven-league boots of his Schlemihl? How could that global conviviality have been shaped if a world-spanning, transareal conception that did not distinguish between an 'us' and the 'rest of the world' had held the upper hand in that Berlin debate about the New World and had not become one of the many buried traditions of eighteenth and nineteenth century Berlin? Many questions, certainly, but also many answers which in turn raises new questions.

Prussia as a nation-state, as a cultural state, as a military power: beyond these often one-dimensional ideas, the book presented here unfolds the image of a multi-perspective Prussia in different views. These are figures of people to whom Prussia meant something, but who did not—or only temporarily—become significant for the historical coming into being of this vanished country. Figures that reflect a piece of the future that should not have been there; figures that represent a piece of the future, that also reflect from the past future into our future and hold up to us the mirror of their time, their thinking, their possibilities.

They are individual figures, it may be, and yet they stand for representative developments of their time with their works. From Anton Wilhelm Amo, the first black philosopher to matriculate at a Prussian university, to Frederick the Great's projection of the Prussian polity onto New Spain and the reign of Moctezuma, to the Amsterdam-born philosopher Cornelius de Pauw, who published his works in Berlin in French and fueled the worldwide Berlin debate on the New World, from the Jewish salon of Rahel Levin Varnhagen to Heinrich von Kleist's imagination of the Haitian Revolution to Adelbert von Chamisso and Alexander von Humboldt, who was not considered a 'true' Prussian: Buried traditions of a history that have been expatriated from the common image of Prussia come alive. A collection, then, of figures who were not 'true' Prussians? Truly not! The volume tells of mobile Prussians whose relationships arrange themselves into a mobile Prussia that we can walk.

The future of Prussia? It is conceivable only as a mobile of ever new configurations. The integration of Jews as well as blacks, the critical approach to a Machiavellian policy of Frederick the Great up to a foundation of colonial actions in the philosophy of Hegel following Cornelius de Pauw, respect for life in between, be it that of Adelbert von Chamisso between Prussia and France or of Alexander von Humboldt between Europe and the Americas: The following chapters sketch a Prussia that needs to be reconsidered not in its becoming-historical, but in its history of becoming… and can still give us today so much.

If, then, a sequence of views of Prussia is to be unfolded here from the perspective of the history of movement, then the mobile aspects of a social, cultural and, in a broad sense, political polity, which—the author readily admits—has lost none of its power to fascinate, are decisive. Any attempt to immobilize this mobile also runs the risk of curtailing a diversity whose qualities of difference constitute the very subject of Prussia. Prussia as a fascination? The fact that simple fixations of what constitutes a Prussian identity have something dangerous about them, something that is still latent, should not and cannot be concealed here. But it is by no means the central focus of the following pages.

The image of Prussia that presents itself to us is consequently a cubist one: from its vectorial knitting pattern, unsteady movements constantly run through this image and structure without ever coming to a standstill. The two salons of Rahel, with their impressive, if transient and temporary, selection of intellectual greats of Prussia, demonstrate as a polylogical community in a very efficient way what such a community could have been.

What a literary mobile can be was impressively demonstrated by the French writer Michel Butor in his experimental text *Mobile.* What possibilities we have to think about Prussia in a different and vectorial way, these pages shall show exemplarily. If the Humboldt Forum in the Berlin Palace, with its historical facades and its rooms, thinks of its own conditions archipelagically and transarchipelagically in the sense of Alexander von Humboldt from the centre of Berlin, in a certain sense from the Spree network of the Museum Island, i.e. if each of its islands is understood as a self-contained island world and at the same time as a multi-connected island world, as a world of islands and archipelagos, then we need have no fears about Prussia and its cultural heritage: Then a wide range of futures will playfully open up to us, which have a chance of not being buried again for a long time. Then, we may hope, we will be able to live and experience Mobile Prussia.

The individual chapters of the present volume emerged at the same time as the discussions surrounding the Humboldt Forum, which is currently under construction, discussions in which the author certainly took part. Ultimately, these discussions were very much about the question of the extent to which a new Berlin is aware of its historical treasures, but also of its historical obligations, and what role the Humboldt brothers should play in the overall concept as well as in the activities of the Humboldt Forum. A Humboldt Forum without the Humboldts? Impossible. But just as impossible a formula as 'Prussia and the World'.

The endless discussions about the developments in the area of the Forum and the debates surrounding its realization were one thing—and challenging.[2] At the same time, however, the related question arose as to what Prussia could and should not have been: in other words, what the potentials, limits and possibilities of a Prussia once were, which were and are significant for the past future as well as for our future today. The result is the present volume, which sought to focus its answers not on the

[2] Cf. e.g. Ette, Ottmar: Das Humboldt-Forum. Where Humboldt is written… A Plea for a Mobile of Cultures. In: *Lettre International* (Berlin) 86 (Autumn 2009), pp. 203–208.

broadest possible range of historical examples, but on representative figures and cases. Only the reading public can judge whether the selection has been successful.

My thanks go to all those who have supported and promoted the idea of Mobile Prussia in the course of the past years. I would especially like to thank Markus Alexander Lenz, who has always been there for me in word and deed and has given me much advice, which I have tried to take to heart. I would like to thank my many students who challenged me in my seminars and lectures and spurred me on with their questions to make this idea a reality. My personal thanks go to my daughter Judith Thamar, who strengthened my belief in the idea of the Mobile in a difficult environment and gave me many a tip. Unforgettable the common paths through Sanssouci: This volume owes much to these paths that branch out: the *senderos que se bifurcan*. From the place of movement Sanssouci these pages were created and hope for a kind audience.

Potsdam, Germany
March 2019

Ottmar Ette

On the 8th of September in the morning the 'Tschirik,' the scunner we saw at sea, entered the harbor. A Prussian from the region of Danzig, Herr Binzemann, was captain of it.—A Prussian who had become captain of a scunner of the Russian-American Company sailing between Unalaschka and Sitcha has endured and experienced many things in the wide world of which one dreams nothing who has got no further in life than, say, from the lower benches of the school to the lectern.
—Adelbert von Chamisso, Reise um die Welt

En attendant ce triomphal retour,
ne fallait-il point reconstituer autour
de soi une nouvelle Europe, effacer le plus
possible de son esprit ce continent qui
nous accueillait mais nous effrayait?
—Michel Butor, Mobile

Contents

Anton Wilhelm Amo or an Enrollment in Prussia 1
The Enrolment 1
Halle or Coming to Speak 6
Wittenberg or the Life (in) Philosophy 11
The Philosopher in the (Academic) Field 16
On Philosophizing Without a Fixed Abode 19
Jena or Philosophizing at the Borders 22
Africa or the Journey to an Unknown Origin 26
Writing-between-Worlds: Philosophizing Without a Fixed Abode 29

Frederick the Great and Cornelius de Pauw: Prussia, Mexico and the (New) World 35
In the Universe of Texts 35
The Power Over Meaning or the Berlin Debate About the New World 40
In the Universe of the Senses 47
The Sense of Power and the Power of the Senses 58

Dom Pernety and the Berlin Debate 67
Travel/Writing 67
Old World/Other World 72
Pernety/de Pauw 76

Alexander von Humboldt or from Revolution to Revolution 85
'Paris croisé'—Revolution and Circulation 85
Ici est un autre 88
The Port of Havana and the Island as a Space of Movement 92
Space as Movement 95
Vector Dimension and TransArea Studies 98
Paris/Berlin 100

Heinrich von Kleist: Saint-Domingue and the Haitian Revolution 103
From the Breakdown of Living Together 103
Colonial Island Worlds 107
Haiti as a Challenge and Paradigm 111
Figural Constructions 116

A Vision of Conviviality 124
Conviviality and Catastrophe 127

Rahel Levin Varnhagen or the Mobile of Prussia 135
'But Everything Is Over!' 135
In the Salon of Rahel Levin 139
Transitions 144
In the Salon of Rahel Varnhagen 147
Sociability, Communication and Conviviality 151

Wilhelm and Alexander von Humboldt or Humboldtian Science 155
'Everything Is Interaction' 155
The Languages of Science 158
Commonality of Concepts, Commonality of Understanding 161
On Globalism: Prussia and the World 166

From Georg Forster to Adelbert von Chamisso: A Voyage Round the World 171
Georg Forster (1754–1794) 171
  World History and Vectopia 171
  Vectopia and Continued Life 177
Alexander von Humboldt (1769–1859) 183
  Vectopia and Fulfilled Life 183
  Vectopi and Survival 191
Adelbert von Chamisso (1781–1838) 195
  Vectopia of World Experience 195
  Knowledge on the Threshold of Survival 200

Selective Bibliography 209

Anton Wilhelm Amo or an Enrollment in Prussia

The Enrolment[1]

On June 9, 1727, a young man entered the register of the preußischen Universität Halle (Prussian University of Halle) under the serial number 488, a man who wrote the history of philosophy without having really entered the history of philosophy as a permanent fixture. The entry[2] in his hand records that the new student at the university,[3] founded only in 1694, i.e. a few decades earlier, came from Africa, 'from Axim in Guinea', and bore the name Anton Wilhelm Amo. Who was this young man?

With this enrolment, Amo "became the first and, for over 220 years, the last person of African origin to study philosophy and law at the Universität Halle (University of Halle)."[4] At the instigation of his sponsors, he was obviously exempted from paying matriculation fees. This regular enrolment marked the beginning of an academic career, of which the internationally famous anthropologist and craniologist Johann Friedrich Blumenbach, who taught at the Universität Göttingen (University of Göttingen), recorded as late as 1790:

> Our Seel. Hollmann, while he was still a professor in Wittenberg, had in 1734 creeded the negro Ant. Wilh. Amo to the D. of World Wisdom, who has shown himself admirably both in writings and as a doctor, and of whom I have two treatises before me, one of which, in

[1] **I sincerely thank the *Bayreuth Academy of Advanced African Studies*, and Susan Arndt in particular, for supporting my research on Anton Wilhelm Amo during my time as a Fellow of the Bayreuth Academy.**

[2] See Fig. 7 in Brentjes, Burchard: *Anton Wilhelm Amo. Der schwarze Philosoph in Halle.* Leipzig: Koehler & Amelang 1976, p. 32/33.

[3] See Abraham, William E.: The Life and Times of Anton Wilhelm Amo, the First African (Black) Philosopher in Europe. In: Asante, Molefi Kete/Abary, Abu S. (eds.): *African Intellectual Heritage. A Book of Sources.* Philadelphia: Temple University Press 1996, p. 429.

[4] Arndt, Susan: *Die 100 wichtigsten Fragen: Rassismus.* Munich: C.H. Beck 2012, p. 143. See also regarding the history of racism in Germany. In: Hund, Wulf D. (ed): Wie die Deutschen weiß wurden. Kleine (Heimat) Geschichte des Rassismus. Stuttgart: Metzler 2017.

O. Ette, *Mobile Prussia*, https://doi.org/10.1007/978-3-476-05839-3_1

particular, reveals much unexpected and well-digested knowledge in the best physiological works of that time. In a report of Amo's life, which was printed on this occasion in the name of the academic Concilii, great praise is given to his exceptional integrity, as well as his abilities, his diligence and his erudition. It is said, for example, of his philosophical lectures excussis tam veterum quam novorum placitis, optima quaeque selegit, selecta enucleate ac dilucide interpretatus est, and so on.[5]

This digression of the great Blumenbach in his *Beyträge zur Naturgeschichte* on a "Negro" named Anton Wilhelm Amo expresses the astonishment, but also the recognition of the man who not only worked as one of the most influential scientific teachers (of a Georg Forster or Wilhelm and Alexander von Humboldt) in Göttingen at the end of the eighteenth century but must also be considered one of the decisive inventors of racial categories and racial hierarchies.[6] The recognition of the philosopher, who was widely known for his black skin[7] and who—as Blumenbach explained elsewhere—subsequently went to the Prussian court in Berlin as a "court councillor",[8] may seem surprising, especially from the mouth of a scientist who promoted racial thinking; and yet the name of Anton Wilhelm Amo probably remained largely unknown to the nineteenth century in wide circles.

But the silence surrounding this Anton Wilhelm Amo was and is also eloquent. For it is high time to *perceive* the dimensions of his life and thought, which are as characteristic of his epoch as they are groundbreaking for the future. We have to rethink his life and thoughts in a broader sense beyond all specialism and to raise them into the public consciousness.

The fascinating figure of Anton Wilhelm Amo demonstrates to us the complexity with which the contradictory period of the early Enlightenment is to be viewed and the extent to which it is long overdue to understand the Enlightenment in its entirety as a transareal,[9] worldwide network linking the most diverse cultural areas. The reference to the unnamed philosopher from Guinea in Francisco Javier Clavijero's *Alten Geschichte Mexicos* threw a sharp light on this:

If a philosopher from Guinea were to undertake a work modelled on Cornelius de Pauw's and entitled *Philosophical Investigations into Europeans*, he could use the same argument

[5] Blumenbach, Johann Friedrich: *Beyträge zur Naturgeschichte.* 2 vols. 1 part. Göttingen: Johann Christian Dieterich 1790, p. 98 f.

[6] See, among others, Painter, Nell Irvin: *The History of White People.* New York/London: W.W. Norton 2010, pp. 72–90.

[7] See this formulation in the essay by Damis, Christine: Le philosophe connu pour sa peau noire: Anton Wilhelm Amo. In: *Rue Descartes: Collège international de philosophie* (Paris) (juin 2002), pp. 115–127.

[8] Blumenbach, Johann Friedrich: Von den Negern. Einige naturhistorische Bemerkungen bey Gelegenheit einer Schweizerreise. In: *Magazin für das Neueste aus der Physik und Naturgeschichte* (Gotha) IV, 3 (1787), p. 9; see on this Lochner, Norbert: Anton Wilhelm Amo. A Ghana Scholar in Eighteenth Century Germany. In: *Transactions of the Historical Society of Ghana* (Achimota) III, 3 (1958), p. 178.

[9] See on this term and the history of globalization also taken as a basis in the present study Ette, Ottmar: *TransArea. Eine literarische Globalisierungsgeschichte.* Berlin/Boston: de Gruyter 2012.

as de Pauw to prove, for example, the harmfulness of the climate of Europe and the merits of the climate of Africa.[10]

It can be assumed that the New Spanish philosopher Francisco Javier Clavijero, who grew up in what later became Mexico and who was well aware of the writings of Leo Africanus, knew of the existence of Anton Wilhelm Amo in his *Alten Geschichte Mexicos*, first published in 1780 in four volumes. Not without reason, he brings the philosopher Amo, who is not mentioned by name, into a direct connection with the Berlin debate on the New World, which is treated separately in this volume and spans the globe. Here, from a New Spanish perspective, Anton Wilhelm Amo becomes an exemplary party in a debate that—as we shall see—emanated from Prussia's capital, reached worldwide dimensions, but is today largely forgotten, buried, in Prussia and Germany.

Consequently, there are good reasons for memorizing the significant name of this young student at the Universität von Halle an der Saale, especially since this name[11] is composed of very different components, which basically allow us an already differentiated access to this personality, who was as important for the African as for the European history of ideas. For even if the long history of the reception of this philosopher and jurist was able to develop only very slowly, not only in Germany but also internationally: From today's perspective, the life and philosophical development of Anton Wilhelm Amo illuminate aspects that are of great importance for an understanding of not only the European Enlightenment as well as the philosophical discourse of (also not only European) modernity.[12]

Anton Wilhelm Amo is certainly not a key figure for Enlightenment thought; and yet he is able to provide us with an important key for a different, more adequate understanding of Enlightenment and modernity. For in this deported slave and revered doctor of philosophy, in this 'test object' of gracious princes and self-confident thinkers of the early Enlightenment in Prussia, there flashes something of what might be called the buried history of European modernity(ies), a history that reached its first genuine (if long-suppressed) climax in the globally received Berlin debate about the New World[13] in the last third of the eighteenth century. It was the

[10] Clavijero, Francisco Javier: *Historia Antigua de México.* Prólogo de Mariano Cuevas. Edición del original escrito en castellano por el autor. México: Editorial Porrúa[7] 1982, p. 462. All translations into German in this volume, unless otherwise indicated, are by the author (O.E.).

[11] On the problem of the surname Amo, see among others Mabe, Jacob Emmanuel: *Anton Wilhelm Amo interkulturell gelesen.* Nordhausen: Traugott Bautz 2007, p. 16 f. The often speculative dimensions of much research on the surname of the abducted boy can hardly be overlooked in the Amo literature.

[12] See Habermas, Jürgen: *Der philosophische Diskurs der Moderne. Zwölf Vorlesungen.* Frankfurt a. M.: Suhrkamp 1985.

[13] See Ette, Ottmar: Von Rousseau und Diderot zu Pernety und de Pauw: Die Berliner Debatte um die Neue Welt. In: Dill, Hans-Otto (ed.): *Jean-Jacques Rousseau zwischen Aufklärung und Moderne.* Proceedings of the Rousseau Conference of the Leibniz-Sozietät der Wissenschaften (Leibniz Society of Sciences and Humanities) in Berlin on December 13, 2012 on the occasion of his 300th birthday on June 28, 2012 in the Rathaus Berlin-Mitte. Berlin: Leibniz-Sozietät der Wissenschaften (= *Sitzungsberichte der Leibniz-Sozietät der Wissenschaften* 117) 2013,

question of how we want to live together in difference and peace, and how a Europe at odds with itself and yet always pursuing similar interests can and should define itself in a global context.

But how can one imagine the history of Anton Wilhelm Amo? About twenty years before his matriculation, on July 29, 1707, the boy had been baptized Protestant in the castle chapel of Salzthal in Wolfenbüttel, which is the first written record of his life, which was deeply marked by the sharp contradictions of his time. For the man born at the turn of the eighteenth century in present-day Ghana, he probably belonged to the West African Nzema and thus to the ethnic group of the Akan,[14] and initially grew up in his homeland. He had been torn from his parents and enslaved, after which he was probably taken to Amsterdam in 1704 by the Dutch West India Company, a journey that almost certainly took him aboard a Dutch slave or merchant ship, first to the Caribbean (where his brother was later forced to work as a slave in the colony of Surinam) and from there to western Europe. An unusual path? Not for the myriad of victims of the European *Slave Trade* across the *Black Atlantic*.[15]

As a 'gift'[16] the boy became the property of an artful and refined duke. The fact that the representatives of the so powerful Dutch trading company could do this undoubtedly points to the fact that it was a matter of a slave, over whom one naturally believed to be able to dispose of freely. Indeed, there is (as already mentioned) testimony and circumstantial evidence to suggest that at least one of the boy's brothers was taken as a slave to Surinam[17] and thus brought into the slave system of the Dutch Caribbean, the Dutch *West Indies*.

Incidentally, there is much to suggest that the Dukes of Brunswick-Wolfenbüttel were involved in the slave trade in many ways.[18] In this respect, they were an exception neither on the German nor on the European level. The physical transfer of the young West African from the institution of slavery, anchored in the European or

pp. 111–130; on the perspectival nature of this debate with a view to the nineteenth century, see ders: *Alexander von Humboldt und die Globalisierung. Das Mobile des Wissens*. Frankfurt a. M./ Leipzig: Insel 2009.

[14] On this quite controversial affiliation see Firla, Monika: Anton Wilhelm Amo (Nzema, Rep. Ghana). Kammermohr—Privatdozent für Philosophie—Wahrsager. In: *Tribus* (Stuttgart) 51 (2002), pp. 55–90.

[15] See the long classic study by Gilroy, Paul: *The Black Atlantic. Modernity and Double Consciousness*. London: Verso 1993; and the volume *Der Black Atlantic*. Edited by the Haus der Kulturen der Welt in collaboration with Tina Campt and Paul Gilroy. Berlin: Haus der Kulturen der Welt 2004.

[16] See Abraham, William E.: Anton Wilhelm Amo. In: Wiredu, Kwasi (ed.): *A Companion to African Philosophy*. Advisory Editors William E. Abraham, Abiola Irele and Ifeanyi A. Menkiti. Blackwell Publishing 2004, p. 191; reservations against this dominant assumption were formulated by Mabe, Jacob Emmanuel: *Anton Wilhelm Amo interkulturell gelesen*, p. 13.

[17] See for an assessment of this fact also Brentjes, Burchard: *Anton Wilhelm Amo. Der schwarze Philosoph in Halle*, p. 30. Reference is also made here to the testimony of the Swiss ship's doctor Gallandat, which will be discussed later.

[18] See Mabe, Jacob Emmanuel: *Anton Wilhelm Amo interkulturell gelesen*, p. 14 f.

Dutch colonial system,[19] to the Duchy of Brunswick-Wolfenbüttel, where nominally slavery did not exist, ultimately involved a continuation of this enslaved condition outside the institution of slavery.

Was that an end to his slavery? By no means. Thus, with regard to Anton Wilhelm Amo, as with many other historical figures, Michael Zeuske speaks with good reason of the fact that it was a matter hereof "enslaved blacks in Europe" within the framework of a "property slavery".[20] Amo had become the property of another person.

Anton Ulrich, the reigning Duke of Brunswick-Wolfenbüttel known for his committed advocacy of the arts and sciences as well as the author of two novels, accepted the "gift" brought to him by the West India Company and together with his son, Wilhelm August, not less inclined to the Enlightenment bestowed the princely first names on the black boy as baptismal names. The already mentioned baptism of the year 1707, which fell into the time before the conversion of the Duke in 1709 to the Catholic faith and whose execution, moreover, contradicts the thesis put forward by William E. Abraham[21] that the boy could come from an already Christian (and thus necessarily baptized) family from a village near Axim in present-day Ghana,[22] thus brings together in the sign of the Christian faith the probably, though not undisputedly from the then Gold Coast originating name Amo[23] with the first names of a European princely house which was quite 'naturally' in the possession of African servants.

Thus, little Anton Wilhelm Amo became part of the black servants who, as at many other royal courts, were 'kept' as an exotic status symbol and a sign of their power, both for their use and for their pleasure. Thus, the life of this Anton Wilhelm Amo, who since the end of his student years began to add the designation 'Afer' or 'Guinea Afer' to his surname Amo, developed from the very beginning in a field of tension that in *geographical* terms was formed by West Africa and the Caribbean, Holland and Germany, a fact that already on this level unmistakably demonstrates the specifically *transareal* dimension of Anton Wilhelm Amo's life, but also his work.

Let us concede at this point: The Dukes of Brunswick-Wolfenbüttel undoubtedly accorded their godson above-average treatment, at least if we compare this with the

---

[19] See the introductory essay 'Slaveries statt Sklaverei: Ein historisch-anthropologischer Essay' in Zeuske, Michael: *Handbuch Geschichte der Sklaverei. Eine Globalgeschichte von den Anfängen bis zur Gegenwart.* Berlin/Boston: de Gruyter 2013, pp. 1–26.

[20] Ibid., p. 517; see also Arndt, Susan: *Die 100 wichtigsten Fragen: Rassismus*, p. 142 f.

[21] See Abraham, William E.: *The Life and Times of Anton Wilhelm Amo, the First African (Black) Philosopher in Europe*, p. 426.

[22] See Brentjes, Burchard: Anton Wilhelm Amo, First African Philosopher in European Universities. In: *Current Anthropology* (Chicago) XVI, 3 (September 1975), p. 443.

[23] Already Wolfram Suchier speculated that it could be a family name from the Gold Coast, but also a naming by Christian love, whereby the little 'Moor' at the court of Anton Ulrich, however, could also have represented Cupid from time to time; see Suchier, Wolfram: A.W. Amo. Ein Mohr als Student und Privatdozent der Philosophie in Halle, Wittenberg und Jena 1727/1740. In: *Akademische Rundschau. Zeitschrift für das gesamte Hochschulwesen und die akademischen Berufsstände* (Leipzig) IV, 9–10 (June–July 1916), pp. 441–448, here p. 443.

living conditions of other black servants and pages at court. Had this not been the case, we would never have heard more of this 'slave without slavery'. This is in no way to deny or even relativize the tremendous tension, strongly influenced by racial thinking, with which Anton Wilhelm Amo was confronted from the beginning of his life. For his masters, for his 'owners', Anton Wilhelm Amo became an experiment, an experimental object on which the educational capacity of 'the' African was to be tested; and by making use of this opportunity he transformed himself from the object of an action performed on him into an acting subject, into the 'master' (span. *amo*) of his fate, which admittedly never allowed him to develop completely freely.

We know little about the period between Anton Wilhelm Amo's baptism and his enrolment at the up-and-coming preußischen Universität Halle (Prussian University of Halle), which at the time was known for its prominent role within the early German Enlightenment. Chamber accounts of the Wolfenbüttel court from the years 1716 and 1717 as well as 1720 and 1721 and a handwritten receipt of Amo for board money received prove repeated payments to the young man and suggest that he was remunerated for services to the court. However, it remains unclear to this day what kind of education Anton Wilhelm Amo enjoyed in Wolfenbüttel, whether he was allowed to attend the in many respects excellent Ritter-Akademie von Wolfenbüttel (Knights' Academy of Wolfenbüttel)[24]—Wolfram Suchier also brought the famous Hallische Waisenhaus into play[25]—or even initially matriculated at the state university of the duchy in Helmstedt between 1721 and 1727.[26]

But none of this could be conclusively proven until today. Did his masters still see in him the slave that had been given to them as a gift, did they see in him a test object with which to conduct experiments, or the chance to provide a humanistic education (understood in the European sense) in the spirit of the early Enlightenment to inhabitants of other parts of the world? However, one may answer these questions: Anton Wilhelm Amo became a student in Prussia.

Halle or Coming to Speak

When Anton Wilhelm Amo came to Halle and thus to one of the centres of the German early Enlightenment, the conflicts between the Pietists and the Rationalists had passed their first major climax there, since the Prussian king, at the instigation of the clerics, had decreed the banishment of the long internationally renowned Enlightenment philosopher Christian Wolff, as whose ardent supporter Amo was long regarded as a young student and later as a lecturer. But even if Carl Günter Ludovici, himself a vehement advocate of Wolff, also described his contemporary

[24] See Abraham, William E.: The Life and Times of Anton Wilhelm Amo, the First African (Black) Philosopher in Europe, p. 429.

[25] See Suchier, Wolfram: A.W. Amo, p. 443.

[26] See ibid., p. 444; Lochner, Norbert: *Anton Wilhelm Amo. A Ghana Scholar in Eighteenth Century Germany*, p. 170 f; and Lochner, Norbert: Anton Wilhelm Amo. In: *Übersee-Rundschau* (Hamburg) 10 (1958), pp. 22–31 and Arndt, Susan: *Die 100 wichtigsten Fragen: Rassismus*, p. 143.

Amo as "a distinguished Wolffian"[27]—a positioning that remains highly controversial to this day[28]—the African-born philosopher cannot be considered a Wolffian, at least in linguistic terms.

For while Christian Wolff was one of those philosophers of the early German Enlightenment who decisively and sustainably advocated the development of German as the language of philosophy in word and deed, Anton Wilhelm Amo chose Latin as the idiom of his academic communication from the very beginning, but also in his later academic development. A rather 'conservative', retrograde act? Or even a silent agreement with the orthodox forces at the university, which was reprimanded by the Prussian ruler?

Not at all. For Amo's long-term choice of language was probably due to various reasons. On the one hand, Latin was still the dominant language of science at German universities (even if it was by no means undisputed), as it still had a high prestige as an internationally understandable scholarly language, especially in Germany. On the other hand, German was undoubtedly still in its infancy in the field of philosophy, consequently in a phase of the constitution and conceptual differentiation, the outcome of which was still open. Latin, however—and this may have been the decisive element for Amo in his choice of language—had already become, at the beginning of the first phase of accelerated globalization, a globalized world language alongside Spanish and Portuguese, in which fundamentally important writings were written and read not only in Europe but also in the non-European colonies, especially in America.

In contrast, German could at best be regarded as a regional language, which, incidentally, in the further course of the eighteenth century, played only a secondary role within the emerging *République des Lettres* of the European Enlightenment—despite its undeniable rise, especially in the field of philosophy. The use of Latin, on the other hand, offered, at least in principle, the chance to be read by a specialized reading public far beyond the borders of the German principalities.

Thus, Amo's choice of language was undoubtedly a sign of a supra-regional ability to communicate that also spanned the borders of Europe, which had to interest the young African beyond his language skills in the field of German, Dutch, French, English, Latin, Greek or Hebrew. Especially in the special boom of globalization under Dutch auspices, the student and doctoral candidate in Halle, Wittenberg and Jena had to be clearly aware that German was not a world language, but Latin—which still fulfilled considerable communication tasks, especially in a global context—very much was. Amo's oral and written use of language, as speaking and

[27] Ludovici, Carl Günter: *Entwurf einer vollständigen Historie der wolffischen Philosophie.* Part III, vol. 1, 3, Leipzig 1738, p. 230. See critically the dissertation by Edeh, Yawovi Emmanuel: *Die Grundlagen der philosophischen Schriften von Amo. In welchem Verhältnis steht Amo zu Christian Wolff, dass man ihn als einen ‚führnehmlichen Wolffianer' bezeichnen kann?* Essen: Die Blaue Eule 2003, p. 11 and *passim.*

[28] The dissertation by Yawovi Emmanuel Edeh just cited makes a full effort to provide scientific evidence that Amo is not to be counted among the Wolffians; see on this also the concluding chapter of his volume in ibid., pp. 144–162. The opposite position can also be found, among others, in Abraham, William E.: *Anton Wilhelm Amo*, p. 195 f.

writing beyond his mother tongue, was in any case marked by a *translingual* language crossing, which from the very beginning represented one of the many far-reaching consequences of his deportation from West Africa.

This had to be of great spiritual significance for a person who had grown up in a (mother) language he had never mastered, to which he could no longer find access from Germany, and which may have been of only local and regional importance. The fact that the surname Amo referred to the lost mother tongue and thus signalled one of the affiliations, while at the same time having an evident relationship with Latin, must have represented an important additional aspect for Anton Wilhelm Amo. In the definitive form of the name Antonius Guilielmus Amo Afer, two externally imposed names—Anton Wilhelm—are thus juxtaposed with two self-chosen names and, at the same time, with two languages and cultures, marking the whole complexity, but also vectoriality, of the young man's situation. Thus, the name of the philosopher forms, as it were, his first and fundamental text, which consciously inscribes itself in different traditions, languages, cultures and designations of origin. In the name, the transreal dimension of this philosopher's life, thought and writing is vital (and visible to all).

At the same time, the writing of the 'black philosopher' is in a multiple sense writing beyond and in the absence of the mother tongue: A gain based on a forcibly enforced lack. Not writing in one's mother tongue was, of course, by no means a novelty or even an exception in the eighteenth century—especially in the context of the emerging *République des Lettres*. Anton Wilhelm Amo, of course, knew that many had to know that he had been deported from a linguistic community not perceived as a culture in the Europe of the early Enlightenment in order to learn and speak the language and languages of his masters. That this consciousness was associated with a special relationship not only to language but to multilingualism, is very evident in his life as well as in his writings. His work on language, which was not solely conceptual, was undoubtedly also due to an attempt to be more than *his master's voice* and to speak more than the language of the colonized object. We can understand this protracted process that Amo underwent as an actual coming to language, in the course of which Amo learned to constitute himself as the subject of his history. A history that was extraordinary because of his educational background and yet at the same time representative of all those People of Colour who had to eke out a living as the property and possession of princes at German courts.

We must, therefore, speak here of a *translingual* field of tension in which the vectorized geographical reference systems sought their adequate linguistic expression, finally finding it in Latin, that 'dead' language from a bygone era which, since the end of the fifteenth century, had once again become a world language and the language of scholars. Amo's work on language is, of course, never monolingual. Even where at first sight monolingual modes of expression are concerned, for example, in academic discourse, the reference to other languages appears again and again in the text—in his Wittenberg dissertation right down to the title of Latin and Greek terminology: *Dissertatio inauguralis philosophica de humanae mentis APATHEIA seu sensionis ac facultatis sentiendi in mente humana absentia et earum in corpore nostro organico ac vivo praestantia.*

Translingual movements in Antonius Guilielmus Amo Afer are thus a consequence of the transareal patterns of experience that he had to go through from childhood on all levels of his existence. At the same time, however, his multilingualism gave him the opportunity to develop a multi-logical way of thinking, a polylogical basic attitude, which can be worked out again and again in his writings. This gives rise to an extremely complex relationship between coming to speak and being spoken to, which characterizes Amo as a subject.

Probably in November 1729, about two and a half years after his matriculation, Amo held his Disputatio on *De iure Maurorum in Europa* at the Universität Halle (University of Halle) under the chairmanship of the renowned jurist, diplomat and specialist in international law, the then chancellor of the university Johann Peter von Ludewig, a public presentation with which he concluded his studies at the Prussian university. Ludewig, who was later also one of the political and diplomatic advisers of the kings of Prussia, may be counted among the most important patrons of Amo at the Universität Halle (University of Halle). He may well have had some influence on the subject of the public disputation.[29]

But the chosen subject from the field of law undoubtedly formed *the* topic for the student, who by now was already experienced and outstanding among his fellow students: It was literally tailor-made for him. For here it was primarily a question of the legal position of those 'Moors' and 'Court Moors', those Africans in Europe who lived in a largely lawless position, as it were as "slaves without slavery,"[30] in Europe. Thus, this jurisprudential disputation is certainly remarkable in the highest degree, not only with regard to the history of international law.[31] It was about nothing less than a determination of rights, but also about the self-determination of the African-born regarding also and especially his rights in Europe. In essence, this public discussion—and this is how the title of the Disputatio could also be translated—was about nothing other than the clarification of the *rights of Africans in Europe*.[32] Here, Amo was negotiating on his behalf. And he had come to the discussion in order to be able to bring up in the legal discourse what was most closely connected with his knowledge of life[33]—as a person of African origin in Prussia.

This first writing of Anton Wilhelm Amo could not be found anywhere and is considered lost since then. It is also still questionable whether it was ever printed. Its topic was explosive. The fact that the disputation in Halle demonstrably caused a great stir, but did not appear even in the historiographical listing of all legal

---

29 See on this the certainly too far going hypotheses of Suchier, Wolfram: A.W. Amo. S. 444; and Lochner, Norbert: *Anton Wilhelm Amo. A Ghana Scholar in Eighteenth Century Germany*, p. 171 f.

30 Zeuske, Michael: *Handbuch Geschichte der Sklaverei. Eine Globalgeschichte von den Anfängen bis zur Gegenwart*. Berlin/Boston: de Gruyter 2013, p. 517.

31 See Mabe, Jacob Emmanuel: *Anton Wilhelm Amo interkulturell gelesen*, p. 20 f.

32 See also Bess, Reginald: A.W. Amo: First Great Black Man of Letters. In: *Journal of Black Studies* (Thousand Oaks) XIX, 4 (June 1989), pp. 387–393, here p. 391.

33 See on this Ette, Ottmar: *ÜberLebenswissen. Die Aufgabe der Philologie*. Berlin: Kulturverlag Kadmos 2004.

disputations,[34] may additionally astonish and not by chance nourish the suspicion that[35] there could have been bad intentions and a hostile hand at play here, in order to eliminate Amo's reflections from the world once and for all. But also the arguments that this writing was never printed cannot be completely dismissed.[36] So, here we are dependent on circumstantial evidence.

William E. Abraham emphasized that Amo had wanted to reach as broad an audience as possible with his reflections and to trigger a reflection on the largely lawless situation of Africans in Europe.[37] If this was indeed his goal, the realization of this intention was probably prevented by the disappearance of his writing (if it ever existed in printed form) at least until our days. In any case, at least according to current knowledge, Anton Wilhelm Amo did not return to the problem he developed in *De iure Maurorum in Europe* in his later publications. Does this mean that when he came to speak he noticed that he was not allowed to bring up everything?

In his explanations, Amo referred back to Roman law as well as to the legal titles of African kings within the Roman Empire. This was certainly necessary, since—as in the area of slavery—Roman law, in his view, still applied everywhere where, for example, Germanic law did not make any more detailed statements or enact any provisions.[38] At the core of Amo's argumentation seem to have been clearly the legal titles granted to the Africans, which opposed any legitimation of slavery.

The fact that the politically so influential von Ludewig, who may be considered neither a friend of Wolff's philosophy nor of its representatives, stood up for the 'Moor in Europe' Anton Wilhelm Amo, also encouraged him on his further way and supported him until his death in 1743, obviously also includes Amo's reflections not only on the abstract legal situation of Africans in general but also on the concrete situation of the young philosopher himself. Everything indicates that Amo could rely on von Ludewig as an advisor and friend right up to his time at the Universität Jena (University of Jena). His princely lords and patrons were increasingly replaced by a figure who, as university chancellor and legal scholar, possessed the subtlest knowledge of the academic field.

At the beginning of Anton Wilhelm Amo's writings, there is thus self-determination and even more self-reflection that allowed him to cleverly thematize and discuss his situation and position, from which he spoke and could speak, and to connect it with his person, with his name. The subject had not only shown himself to be intelligent but at the same time had made an attempt of his own in the matter of self-determined living. A tipping point, however, arose between "freeness" and

---

[34] Thus, Wolfram Suchier points out in his early, pioneering study that even in Christoph Weidlich's list of all legal disputations at Halle from 1789 any reference to Amo's Disputatio is missing; see Suchier, Wolfram: A.W. Amo, p. 444.

[35] See on this Sephocle, Marilyn: Anton Wilhelm Amo. In: *Journal of Black Studies* (Thousand Oaks) XXIII, 2 (December 1992), p. 185 f; and Bess, Reginald: A.W. Amo: First Great Black Man of Letters, p. 390 f.

[36] See Brentjes, Burchard: *Anton Wilhelm Amo. Der schwarze Philosoph in Halle*, p. 37.

[37] See Abraham, William E.: *The Life and Times of Anton Wilhelm Amo*, p. 430.

[38] See also ibid., p. 430 f.

"servitude," especially in view of the "Moors in Europe bought by Christians."[39] This laid the foundations for the acquisition of a multi-perspective philosophical position, which is probably unique in African-European relations within Enlightenment philosophy in the German-speaking world—and not only there.

By subsequently turning to questions of the kind that moved the philosophy(s) of the early German Enlightenment, he secured the sympathy and also the support of many colleagues and some patrons, above all the influential Johann Peter von Ludewig, whose death in 1743 undoubtedly marked an important turning point for Anton Wilhelm Amo. From the reflection of his deportation as well as his transplantation into a context completely different from his country of birth, a specific dimension of an inside *and* outside sensibility emerges in the emerging philosophy of Antonius Guilielmus Amo, which was repeatedly called to consciousness by the addition of the word 'Afer', and which undoubtedly made the investigation of epistemological questions, in particular, appear to be urgent. The African origin of the German philosopher in Europe in no way reduced the other components of his name, in which his coming to language is concentrated and crystallized as in probably no other text.

Wittenberg or the Life (in) Philosophy

At the time when Amo was sent by his ducal patrons, but also 'owners' and masters, to the up-and-coming preußische Universität Halle (Prussian University of Halle), the struggle of the strong pietist faction against the rationalists was still raging there, whose most important head, Christian Wolff, certainly the most original German Enlightenment philosopher between Leibniz and Kant, had had to pay for his philosophical position with immediate banishment from Prussia. Prussia had taken a stand. Nevertheless, there were many professors in Halle who, like Christian Thomasius, who was closely associated with Wolfenbüttel and who, in the field of jurisprudence, advocated a humane criminal law-oriented towards the ideas of the early Enlightenment and who, moreover, had set important accents in the problem of the body and soul, were also known for their open and freedom-loving Enlightenment positions.

Even after 1723, when Christian Wolff had to leave Halle within forty-eight hours on the orders of the Prussian ruler Friedrich Wilhelm I, the often vehement disputes and power struggles continued unabated, in which, in particular, Christian Thomasius (who, however, had already died in September 1728 and thus very early for Amo) and Johann Peter von Ludewig were exposed to repeated hostility.[40] Anton Wilhelm Amo must have understood very early on that scientific and intellectual positions never settle in an abstract space, but are the results of complicated mediations.

---

[39] Quoted from Brentjes, Burchard: *Anton Wilhelm Amo. Der schwarze Philosoph in Halle*, p. 38.

[40] See ibid., p. 34 f.

The constant changes within this field, which was by no means autonomous but directly affected by state and clerical interventions, were certainly also related to Anton Wilhelm Amo's decision to turn his back on the Universität Halle (University of Halle) and thus on Prussia for the time being and to move to the Universität Wittenberg (University of Wittenberg), which was not yet affiliated with the latter at the time. In view of the academic tensions outlined above, this was undoubtedly a fortunate decision, to which Amo certainly owed his happiest time at a German university.[41] At Halle, the conflicts between Wolffians and Pietists continued, with the subsequent departure of several important professors (such as the aforementioned Ludovici) from Prussia. The Universität Halle (University of Halle) had lost its importance.

Amo joined this exodus from Halle, which was not without danger for the continued existence of the university. And in the sense of Bourdieu's field theory,[42] one could certainly assume that leaving Prussia was to a certain extent also tantamount to leaving a subfield of philosophical positional struggles that was specific to Prussia, but also the University of Halle (University of Halle). This opened up a second academic playing field for Amo, even if this was not fundamentally separate from the university there, if only because of its proximity to Halle. But crossing the Prussian border offered additional, if only momentary, protection for his person.

Thus, on September 2, 1730, Amo entered the matriculation register of the Universität Wittenberg (University of Wittenberg), where at that time the Pietists were not yet able to develop a comparable power. Only a few weeks later, on October 17, 1730, he was awarded a doctorate in philosophy and the liberal arts: His university achievements from Halle had been recognized with a rapidity that, from today's perspective, must arouse admiration, perhaps even shame. Amo's departure across the state border to Wittenberg was also a shrewd move in that he now had a first academic degree on which he could now build at another prestigious university.

At the same time, Amo now changed his scientific profile and turned increasingly to medical studies. He, thus, laid the foundations for his later dissertation, which was to be located in the borderland between medicine, anthropology and philosophy. Under the caring guidance of the physicist Martin Gotthelf Loescher, who was one of the moderate mechanists in the field of medicine,[43] the freshly minted Magister worked his way swiftly into the research of the new subjects and gained widespread recognition among professors and students alike in Wittenberg.

It may be said without any exaggeration that the Universität Wittenberg (University of Wittenberg) boasted of its black magister and deliberately placed him at the head of the delegations that received the sovereign on the occasion of a visit to the university town by the new Elector of Saxony, Frederick Augustus III, later

---

[41] See also Brentjes, Burchard: *Anton Wilhelm Amo. Der schwarze Philosoph in Halle*, p. 38 f.

[42] See also Jurt, Joseph: *Das literarische Feld. Das Konzept Pierre Bourdieus in Theorie und Praxis*. Darmstadt: Wissenschaftliche Buchgesellschaft 1995.

[43] Brentjes, Burchard. *Anton Wilhelm Amo. Der schwarze Philosoph in Halle*, p. 41.

King of Poland.[44] Within a few years, Amo had acquired a high prestige in the university town.

On May 10, 1733, so the *Hamburgische Berichte von neuen Gelehrten Sachen* of June 2 of the same year testify, the honorary formations had lined up in the presence of the gracious Saxon Elector for the hereditary homage and honorary testimony. Already a small excerpt from the detailed report may give us a vivid picture of the role Anton Wilhelm Amo took over:

> Mr. M. Amo, an African, stood in the middle, as commander of the whole corps, dressed in black, carrying a primitive stick in his hand, and wearing over his waistcoat a broad white ribbon, on which the Saxon coat of arms was splendidly embroidered with gold and black silk. Saxon coat of arms was splendidly embroidered with gold and black silk mixed in. The two marshals, however, who accompanied him to the college, wore black ribbons over their shirts so that there might be a difference between them as far as the finery was concerned.[45]

The 'black philosopher', who appeared in splendid black regalia, had finally become a showpiece with which one could make state on official occasions. Certainly, one can also read this highlighting of the black magister, who participated in several processions in honor of the Saxon elector and commanded his marshals, as a tipping figure, insofar as the exceptional position of Amo could very quickly turn from inclusion into its opposite of exclusion, as one or the other of the illustrious spectators might feel reminded of the role of the court moors, especially at German princely courts. The prominent position of Amo is therefore to be judged ambivalently.

But Amo's honor was undoubtedly also based on the excellent reputation that the young master of philosophy and liberal arts had earned at the Universität Wittenberg (University of Wittenberg) in the course of the previous years. And also in other respects, the "African" seems to have been integrated into the university life, as the city archives of Wittenberg still keep lists in which—as the city archivist Kühne found out[46]—the defaulting revelers were recorded—among them none other than Magister Amo.

But this budding philosopher, widely recognized and respected in Wittenberg, also entered another list: That of doctoral students preparing their dissertations and practicing the discourse of schooling in German philosophy in the era of the early Enlightenment. If Anton Wilhelm Amo had dealt with a legal problem of vital interest to him in Halle, his cunning in Wittenberg consisted in now pursuing an epistemological question dealing with the relationship between body and soul, with the body-soul problem. With this topic, which undoubtedly belonged to the important, perhaps even central, in any case strongly contested questions of the philosophy of that time, Amo was concerned with a highly innovative view of the relationship

---

[44] See Lochner, Norbert: *Anton Wilhelm Amo: A Ghana Scholar in Eighteenth Century Germany*, p. 174 f.

[45] *Hamburgische Berichte von neuen Gelehrten Sachen* (Hamburg) XLIX (June 2, 1733), p. 366.

[46] This draws attention to Brentjes, Burchard: *Anton Wilhelm Amo. Der schwarze Philosoph in Halle*, p. 46.

between body and soul, whereby the self-imposed task consisted in assigning the attribution of sensations to the body or the soul. Only at first glance does this question appear to be a purely academic problem. But what is so exciting about it?

In his twenty-four-page inaugural dissertation *De humanae mentis apatheia* (the full title of which should be translated as *The Apatheia of the Human Soul, or On the Absence of Sensation and the Capacity for Sensation in the Human Soul and the Presence of Both in Our Organic Living Body*), Amo pursued this question in Latin in a highly systematic and concise manner. The lexeme *vivo*, which is central to the work, is already found in the title. This eminently important lexeme of life is repeated by Amo in his tightly and clearly structured dissertation, after the enumeration of two negative theses in the third, now affirmative thesis in the title formulation, right on the first page, where it speaks of "*our organic living body*".[47]

There are good reasons for recognizing this inaugural dissertation, at least in a broad sense, as a study oriented towards the life sciences. For while *De iure Maurorum in Europa* was concerned with a jurisprudential question integrating the own life knowledge of those born in Africa and deported to Europe, the dissertation pursued the question of life from the perspective of body-soul dualism. The intertextual interlocutors in this study were not only authors of medical or physiological treatises, but also philosophers of the rank of John Locke or René Descartes, with whom the black German philosopher critically engaged.

The inaugural dissertation presented in Wittenberg in 1734, publicly defended under the chairmanship of Martin Gotthelf Loescher and published in print in the same year, assumes that the human body "as something very perfect from various vital and animated organs was first artfully manufactured by the Creator and also reproduced by procreation".[48] Here the body is understood as a *living* entity, produced by a creator, begotten by human beings and driven forward by a life force (*vis vitalis*), whereby Amo bases himself here on writing by Christian Vater that he cites.

---

[47] I use here the German edition of *De humanae mentis apatheia* in: Antonius Guilielmus Amo Afer from Axim in Ghana: *Übersetzung seiner Werke.* [Trans. A. Blaschka?] Halle (Saale): Martin Luther University Halle-Wittenberg 1965, here p. 71. This translation is not without problems, as already the updating indication of the author (from Ghana, not from Guinea as in the original) shows. The original edition in Latin was published in Wittenberg in 1734 by Schlomacher under the title *Dissertatio inauguralis philosophica de humanae mentis APATHEIA seu sensionis ac facultatis sentiendi in mente humana absentia et earum in corpore nostro organico ac vivo praestantia.* A somewhat different translation can be found in the appendix to Brentjes, Burchard: *Anton Wilhelm Amo. Der schwarze Philosoph in Halle*, pp. 87–108. The original quotation in Latin reads: "corpori nostro organico et uiuo" (here as in the following quoted according to the facsimile edition of the *Dissertatio inauguralis philosophica de humanae mentis APATHEIA seu sensionis ac facultatis sentiendi in mente humana absentia et earum in corpore nostro organico ac vivo praestantia.* Philosophical Inaugural Dissertation, Wittenberg 1734, facsimile edition). Halle (Saale: Martin Luther University Halle-Wittenberg 1978, p. 3).

[48] Antonius Guilielmus Amo Afer: *Übersetzung seiner Werke*, p. 76. In the original Latin: "ad corpus quod adtinet est: elegantissimum e diuersibus organisuitablibus et animalibus a creatore primum fabrefactum, et de hinc quoque per generationem propagatum" (*De humanae mentis apatheia*, p. 9).

According to Amo, all things have an inherent life force. The central concept of sensation is thereby connected with that of life:

> Everything that feels, that lives. Everything that lives takes in nourishment. Everything that lives and takes in nourishment grows. Everything that is of this kind is finally resolved into its primordial substances (prima principia). Everything that is resolved into its primordial substances is composed of primordial substances (principiatum). Everything that is composed of primordial substances has its constitutive parts. Everything that is of this kind is a divisible body. Therefore, if the human soul feels, it follows that it is a divisible body.[49]

Anton Wilhelm Amo uses all the sharpness of his mind to prove, with the help of his logic and many a syllogism, what he has already clearly formulated as an assertion in the first part of his work, right at the beginning of his remarks: Namely, that "every spirit stands outside of every suffering"[50]—and he affirms that if something is material, "that it cannot be immaterial".[51] In his radical body-mind dualism, he consequently sharply separates the two realms.

If we follow the main thesis of the inaugural dissertation, then the human soul and spirit is separated from everything material, sensual and corporeal, and is, as it were, 'purified' (and thus withdrawn from the corporeal, i.e. *abstract*). One could very well associate Amo's body-mind dualism, which may well be assigned to a moderate mechanism[52] and which contains a quite clear critique of René Descartes's positions,[53] with an analogous detachment that protects the soul, the spiritual from all that is physical and thus detaches it from all physical constitution.

The freedom of the mind is undoubtedly one of the central theorems of Anton Wilhelm Amo's philosophy. And this central conviction is likely to be fed by sources and, above all, a knowledge of life that is highly specific to Amo and are rooted in his role as a person of black skin color among whites, a role that he could not circumvent from a practical point of view. In this respect, his philosophy conveys a great deal of autobiography, not only with regard to his legal situation but also to the central concepts of contemporary philosophical debates.

One could also derive from Amo's reflections the conclusion that the immateriality of the spirit at the same time implies a freedom that makes *apatheia* possible in the positive sense, that is, equanimity and peace of mind, entirely in the tradition of

---

[49] Antonius Guilielmus Amo Afer: *Übersetzung seiner Werke*, p. 101 f. In the original Latin: "Quidquid sentit, illud uiuit, quidquid uiuit nutritur, quidquid uiuit et nutritur augmentatur, quidquid huius modi est, tandem in sua Prima principia resoluitur, quidquid in sua prima principia resoluitur, est principiatum, omne principiatum habet suas partes constitutiuas, quidquid eius modi est, est corpus diuisibile si igitur mens humana sentit, sequitur quod sit corpus diuisibile" (*De humanae mentis apatheia*, p. 15).

[50] Antonius Guilielmus Amo Afer: *Übersetzung seiner Werke*, p. 73. In the original Latin: "Omnem spiritum esse extra omnem passionem" (*De humanae mentis apatheia*, p. 5).

[51] Antonius Guilielmus Amo Afer: *Übersetzung seiner Werke*, p. 74. In the original Latin: "Si aliquid est immateriale sequitur quod materiale esse nequeat" (*De humanae mentis apatheia*, p. 6).

[52] See Wiredu, Kwasi: Amo's Critique of Descartes' Philosophy of Mind, in Wiredu, Kwasi (ed.): *A Companion to African Philosophy*, pp. 200–206.

[53] See Brentjes, Burchard: *Anton Wilhelm Amo. Der schwarze Philosoph in Halle*, p. 46.

the Stoa. But this form of *apatheia*, of unshakable peace of mind, to whose exponent Anton Wilhelm Amo referred in his later *Tractatus de arte sobrie et accurate philosophandi*,[54] can also be a weapon in the philosophical struggle, especially when thought, when the activity of the mind, is not tied back to and dependent on the material, the physical dimension of the body. The peace of mind of this 'apathy' includes in itself greatness of soul that can also lay claim to being able to argue on an equal footing as a matter of course, regardless of the particular consideration of the body. In Amo's *Tractatus*, to which we shall return later, *apatheia* also appears in the title, since Amo is concerned here with philosophizing not only under the sign of the accurate and precise but also of the *sobrie*, the sober, unshakable and calm of soul. It is impressive how much strength and power, but also self-control and resistance, speak from the reflections of the young, distinguished philosopher.

With *De humanae mentis apatheia*, Anton Wilhelm Amo not only proved the success of his academic socialization at German universities, but also his talent for combining the sharpness of his independent thinking with insight into the rules of the academic field. In the philosophical reflection of life knowledge, Amo gave philosophically expressive proof of his survival knowledge[55] in the field of philosophy.

These were undoubtedly happy years for Amo as well as for Amo's philosophy. In the same month of the public defense of his dissertation, Anton Wilhelm Amo was appointed and admitted as Magister legens at the Universität Wittenberg (University of Wittenberg) on April 16, 1734. He could hope to make a career as a scientist in Germany.

The Philosopher in the (Academic) Field

With the defense of the dissertation in the following year, all doors in the academic field were, it seemed, open for Amo. In April 1734, Martin Gotthelf Loescher, as chairman of Amo's disputation, once again noted the young man's outstanding abilities and referred right at the beginning of his remarks to "Africa and its country Guinea, which lies quite far from us and which was once called the Gold Coast by the Europeans because of its immensely rich deposits of gold."[56] It is certainly no coincidence that also in the two last-mentioned documents the most illustrious patrons and sponsors, but also lords of Anton Wilhelm Amo was mentioned first. If here mostly also grace, kindness and paternal clemency are spoken of, this

[54] See on the tradition of the Stoa in Amo's thinking as well as on his attempt at a synthesis between Epictetus and Christianity, among others Firla, Monika: *Ein Jenaer Stammbucheintrag des schwarzen Philosophen Anton Wilhelm Amo aus dem Jahr 1746*. Stuttgart: AfriTüDe Geschichtswerkstatt 2012, p. 9 f and 21.

[55] See on this Ette, Ottmar: *ÜberLebenswissen. Die Aufgabe der Philologie* (2004).

[56] Reprinted in German translation in Antonius Guilielmus Amo: *Übersetzung seiner Werke*, p. 87. In the original Latin: "Africam & ejusdem longissime a nobis dissitam regionem Guineam, olim ora aurea, ob copiosissimum auri proventum ab Europaeis appellatam" (*De humanae mentis apatheia*, p. 23).

nevertheless throws a significant light on the dependent situation in which the 'black philosopher' was very well known. For his owners and possessors discreetly brought themselves to mind. Yet Anton Wilhelm Amo had every reason to believe that the university career he aspired to was within his grasp and to reach out for it.

But soon the prospects for the aspiring philosopher with a doctorate darkened. Not only because in 1735 the last of his patrons among the Dukes of Brunswick-Wolfenbüttel, Duke Ludwig Rudolf, died. Not only because Löscher, Amo's friend and teacher, also died in the same year.[57] The decisive factor for the change of the situation was certainly the fact that now also in Wittenberg the fight against all those who were considered followers of Wolff's philosophy was waged with increasing vehemence, and in orthodox circles, even a "pulpit ban for all followers of Wolff-Leibniz's philosophy was discussed" in all seriousness.[58]

Thus, however, the academic field for Amo, who in Wittenberg repeatedly added the geographical suffix 'African from Guinea' to his name, was transformed once again and more clearly than ever before into a quasi-military field of conflict, which very much seriously threatened his material survival as a philosopher. For Anton Wilhelm Amo was no longer a student: He was teaching himself and had already found his first pupils. Although the subject of his experiments had become a show-piece, Amo evidently now increasingly appeared to some as a threat. Friends and supporters who could protect him became rare. Amo was increasingly on his own.

Anton Wilhelm Amo took action: He returned to Prussia in order to make a second attempt at the Universität Halle (University of Halle), in an atmosphere that was somewhat less heated at the time, to further develop his independent philosophy, based on his knowledge of life and knowledge for survival, within the framework of the early Enlightenment universe of discourse. Prussia still seemed tempting to him here. He knew that if he wanted to take his still-present chance to continue his dream of an academic career at German universities, he would have to further qualify himself and produce more extensive writing as a philosopher. There was no reason at this point to abruptly abandon this goal, which he had now been pursuing for years. So he placed his hopes once again on Prussia.

If Amo began his university career in the Prussian sphere of power at a time[59] when the 'Brandenburg-African-American Company', which had been established since the early 1680s, continued to ship deported African slaves from the so-called Gold Coast to the Caribbean under the Brandenburg or Prussian flag, this fact forms a highly contradictory background for a career that was to be regarded as

[57] See Brentjes, Burchard: *Anton Wilhelm Amo. Der schwarze Philosoph in Halle*, p. 47.

[58] Ibid.

[59] See Lennert, Gernot: Kolonisationsversuche Brandenburgs, Preußens und des Deutschen Reiches in der Karibik, pp. 11–23; and Weindl, Andrea: The Slave Trade of Norther Germany from the Seventeenth to the Nineteenth Centuries. In: Eltis, David/Richardson, David (eds.): *Extending the Frontiers: Essays on the New Transatlantic Slave Trade Database*. New Haven: Yale University Press 2008, pp. 250–271. On the source situation, see Jones, Adam: *German Sources for West African History*. Wiesbaden: Franz Steiner 1983; as well as Jones, Adam: *Brandenburg Sources for West African History 1680–1700*. Stuttgart: Franz Steiner 1985.

emancipatory from the most diverse perspectives. Anton Wilhelm Amo stood for this emancipation with his curriculum vitae.

It was not until 1738, after the deportation of at least 30,000 Africans, that the 'Soldier King' liquidated the trading company, which had long been in difficulties due to bad business, and whose dissolution thus came before the death of King Frederick William I on May 31, 1740 and before the accession of his son Frederick II of Prussia.[60]

In this Prussia, then, whose colonialist dreams seemed to be over after the first third of the eighteenth century, the increasingly registered presence of the aspiring and successful philosopher, now well over thirty years old, looks like the inverse representation of a society that had failed in its colonialist ambitions. Against the background of Prussia's colonial dreams and the country's involvement in the transatlantic slave trade, the figure of Anton Wilhelm Amo seems almost provocative.

Their undoubtedly greatest philosopher after Leibniz in the German-speaking world, Immanuel Kant, in whose work, unlike that of his contemporary Blumenbach, there seems to be no reference whatsoever to Amo, was to become decades later an important legitimizing figure not only of racial thinking but of racist ideas, which the philosopher from Königsberg set down in his writings precisely with regard to the black population in Africa and elsewhere.[61] Should he really never have heard anything in his Prussia about the German philosopher from Africa who taught in Halle, Wittenberg and Jena? This is anything but likely: Kant was one of those who literally hushed up Anton Wilhelm Amo both during his lifetime and long after his death. Thus, within just a few years, Amo was to form one of the many buried traditions in Prussia and the German-speaking world that the present volume attempts to uncover.

Conversely, the history of Brandenburg-Prussian colonialism must certainly have been very familiar to Anton Wilhelm Amo, even if no statements on this subject can be found in his writings. How could he have been unaware of these activities, which, moreover, were located in a geographical area that included the region of the African coastal zone divided between various European colonial powers, in which he had been born? Amo was undoubtedly aware of the anti-colonial implications and explosive power of his sheer existence, and even more so of his academic brilliance.

In the dean's book of the Faculty of Philosophy at the Universität Halle (University of Halle), a note can be found confirming Anton Wilhelm Amo's

[60] See the following chapter in this volume.

[61] See on this, among others, Arndt, Susan: *Die 100 wichtigsten Fragen: Rassismus*, pp. 61–63; and, against the background of the study of Amo Gordon, Lewis R.: *An Introduction to Africana Philosophy*. Cambridge: Cambridge University Press 2008, p. 37 f; Mougnol, Simon: *Amo Afer. Un Noir, professeur d'université, en Allemagne au XVIIIe siècle*. Paris: L'Harmattan 2007, pp. 25–27; as well as the statements in the publication by Schilling, Georg, "*Who was Anton Wilhelm Amo?*"—"*Remapping the Jungle*" *vs.* "*Vom faulen Holze lebend*"*? An attempt against forgetting and* "*white*"*-washing, sad, racist* "*(a)e(sthe)tic(al)*" *texts of so called German* "*prime fathers*" et al. Munich: GRIN 2009, p. 24 and *passim*.

admission as a lecturer at his alma mater on July 21, 1736.[62] An important role for the philosopher within the changed situation, which was again more open to the Wolffians, may again have been played by the chancellor of the university, Johann Peter von Ludewig, with whom Amo certainly kept in touch from nearby Wittenberg. Thus, under von Ludewig's protection in Halle, Amo was able to continue his activity as a lecturer, which—as we have seen—he had already carried out with apparently such great success in Wittenberg. The academic knew only too well, due to his oppressive financial circumstances, which he mentioned very well in his application to the Faculty of Philosophy,[63] in what a contested field he was moving.

In this stage of his life, however, the philosopher had to be primarily concerned with decisively expanding his philosophical oeuvre in the narrower sense. Alongside cleverness, perseverance was undoubtedly one of the character traits and qualities of an ambitious university teacher.

And so in 1738—just in that year when the Prussian crown withdrew from the transatlantic slave trade—his actual philosophical magnum opus could appear in Halle: *Antonii Guilielmi Amo Guinea-Afri, philosophiae et artium liberalium magistri, Tractatus de arte sobrie et accurate philosophandi*. Neither in the run-up to nor in the aftermath of its publication, however, does this work seem to have been spared those disputes that had already urged, indeed forced, Amo, to move to Wittenberg and may have prompted him once again to leave the university town of Halle in the following year, 1739. The former showpiece was increasingly put on the defensive and saw his chances of being able to crown his brilliant academic career in Prussia, Saxony, or other German countries with a permanent professorship dwindle.

On Philosophizing Without a Fixed Abode

That the *Tractatus de arte sobrie et accurate philosophandi*[64] could hardly be received positively in the camp of the Pietists is already evident in the paratextual section of this work. There, in his extensive dedication, dated July 6, 1737, the author thanks not only Johann Peter von Ludewig, but also other representatives of a rationalist, not subject to ecclesiastical constraints, and clearly anti-Pietist[65]

---

[62] This note is printed in Brentjes, Burchard: *Anton Wilhelm Amo. Der schwarze Philosoph in Halle*, facsimile zw. p. 47/48.

[63] See Brentjes, Burchard: *Anton Wilhelm Amo. Der schwarze Philosoph in Halle*, p. 50.

[64] Amo, Anton Wilhelm: *Tractatus de arte sobrie et accurate philosophandi.* Halle: Kitler 1738; I revert to the translation into German under the title Traktat von der Kunst nüchtern und sorgfältig zu philosophieren. In: Antonius Guilielmus Amo Afer: *Übersetzung seiner Werke*, pp. 103–281. The original Latin quotations are taken from the *Tractatus de arte sobrie et accurate philosophandi.* Antonii Guilielmi Amo Guinea-Afri, Philosophiae et artium liberalium magistri. Nendeln: Kraus Reprint 1971.

[65] See also Glötzner, Johannes: *Anton Wilhelm Amo. Ein Philosoph aus Afrika im Deutschland des 18. Jahrhunderts*. Lecture given on the occasion of the 500th anniversary of the Universität

understanding of philosophy, most eloquently, concluding his tokens of gratitude and homage with a memorable modesty topos:

> Therefore, will you kindly receive these attempts of mine, of whatever kind, to make myself deserving of the public, and exercise forbearance, should something human have happened to me in the process; that is what I implore you to do. For the rest, may God keep you, my patrons, in perfect prosperity for both communities, the State and the Republic of Scholars, for both of which you have rendered the best of services and continue to do so up to the present time, and may you live many more years. Farewell and stay in my good graces![66]

A thank you with a farewell? This may come as a surprise. But for the first time, public acting published writing of Anton Wilhelm Amo is no longer 'framed' with the reference to his patrons, lords and 'owners', the Dukes of Brunswick-Wolfenbüttel, but is opened with the naming of those patrons whom Amo—and for this, he might have had good reasons—calls his patrons. This is undoubtedly connected with a clearly marked independence, which leaves the (enslaved) servitude behind and consciously knows itself in the area of own freedom.

In this, Amo identifies two different fields, which he labels 'state' and 'scholars' republic', whereby the autonomy of the latter—as Amo knew very well—could only be a highly relative and always precarious one. But it is unmistakably signalled here that the author explicitly moves freely as an African from Guinea in this scholars' *republic*, this *République des Lettres*, and now wants to freely choose his paths. Especially as an African from Guinea, because he leaves no doubt about this. And this very well in the awareness of the fact how precarious his situation was indeed at that time, not only in the financial area. Does not Anton Wilhelm Amo Afer's claim that philosophy can no longer be the 'property' of Europeans clearly emerge here? Is not the claim clearly made here that one should be allowed to philosophize beyond all racial prejudices and demarcations in order to give philosophy the honor?

While in *De iure Maurorum in Europe*, Amo had used a juridical and in *De humanae mentis apatheia*, an anthropological-medical-philosophical question, in his treatise on philosophizing, in his *Tractatus de arte sobrie et accurate philosophandi*, he moves from the beginning on the genuinely philosophical field of logic. Thereby, in a general part preceding his work, he starts from the question and conceptuality of intention and connects it with its relation to God, but also and especially to the human soul,[67] so that in his *Tractatus* he is basically able to directly resume his inaugural dissertation. The coefficient of motion of Amo's philosophy is high in thematic, representational and methodological respects; but the strong continuities of his art of philosophizing can very well be recognized.

Wittenberg-Halle (University of Wittenberg-Halle) on June 27, 2002 in Halle. Munich: Editionen Huber 2002, p. 14.

[66] Amo, Anton Wilhelm: *Traktat*, p. 106. In the original Latin this passage reads: "Quare vt qualescunque hos meos, de publico bene merendi conatus, serena fronte accipiatis, & quidquid humani passus fuerim ignoscatis, est, quod vehementer rogo. Caeterum Deus *Vos, Maecenates*, rei vtrique, publicae ciuili, ac literariae, de que vtraque optime meriti estis, & adhuc meremini, diutissime sospites seruet, pluresque annos perennare faxit. Valete & fauete!" (*Tractatus*, o. p.).

[67] Amo, Anton Wilhelm: *Traktat*, p. 107.

In this sense, the work also exhibits a similarly clear and concise structure as had already been observed in *De humanae mentis apatheia.* Here, too, Amo's clearly epistemologically oriented line of vision is evident, which as a rule starts from the terminological clarification and then the conceptual unfolding of its respective term. As in *De iure Maurorum in Europa* and *De humanae mentis apatheia*, an optimistic view of man can be discerned here, which, moreover, is unquestionably convinced of the efficacy of thought, of the interpretive power of philosophy. Even if—as already mentioned—Amo was called a Wolffian by Wolffians at the time and Burchard Brentjes (together with many who followed him in this) unconditionally agreed with this assessment in his monograph,[68] one should definitely take note of the fact that especially with regard to the explicit thinking, the power of interpretation of philosophy, Amo is a Wolffian, that just with regard to the explicit intertextual relations Amo very well marked—as Yawovi Emmanuel Edeh[69] showed in the first dissertation dedicated to Amo's thought—a clear distance to Christian Wolff's thought, which was undoubtedly familiar to him (and be it for field tactical reasons). In his *treatise*, Anton Wilhelm Amo presents himself as an independent thinker who knows patrons but no 'pre-thinkers', no actual founders of discourse. He treads his path.

In this context, the limits of church doctrine can always be clearly discerned—not in the sense of a pietistic orthodoxy, but of a claim of Christianity to the discursive setting and framing of all philosophy. The overlapping areas of the fields of philosophy and theology had often become starting points of heated disputes, not least in Halle, so that it may be regarded as highly courageous, perhaps even as audacious or breakneck, when Amo just in this field in Sect. IX of the first chapter of his main work ventured out of cover with a definition of theology. There, after an initial, salvation-historically grounded definition of "*Christian[n] theology*" as a "habitus of intellectual and effective intention, by virtue of which we are concerned with the genuine and solid knowledge of the truth",[70] he abruptly states: "*I say explicitly: the theology of Christians. There is, in fact, besides, a theology of the pagans, of the Turks, etc., further, according to the diversity of peoples.*"[71]

With it, Anton Wilhelm Amo not only questioned Christianity's sole claim to a theology (founded by God himself), but also opened, even if only in a flash, the insight into and the outlook on a diversity of religions as well as a diversity of

[68] Also with regard to the *Tractatus*, see Brentjes, Burchard: *Anton Wilhelm Amo. Der schwarze Philosoph in Halle*, p. 52 f.

[69] See Edeh, Yawovi Emmanuel: *Die Grundlagen der philosophischen Schriften von Amo*, et al. p. 163 f.

[70] Amo, Anton Wilhelm: *Traktat*, p. 117; in the original Latin: "Habitus intellectivae & effectivae intentionis, quo versamur circa veram & solidam veritatis cognitionem" (*Tractatus*, p. 15).

[71] Amo, Anton Wilhelm: *Traktat*, p. 118; in the original Latin: "*Notantur dico, Theologia Christianorum. Alia enim Theologia Gentilium, alia Turcarum, alia & alia pro diversitate Gentium*" (*Tractatus*, p. 15).

cultures, yes, recognized the "*diversity of peoples*"[72] in a formulation without any devaluation of those peoples.

The philosophy of Anton Wilhelm Amo can, therefore, be placed in a line of development that could be described—by no means only in the field of the Literature of the World—as Writing-between-Worlds.[73] With regard to Amo and within the framework of the historically given possibilities, we can very well speak of philosophy without a fixed abode, which began to take shape in his works. For it is not committed to and assignable to a single place, a single perspective, but unfolds on the level of the fields of tension mentioned so far in a profoundly vectorial way. It allows the points of view of others on the same objects to emerge, and since *De iure Maurorum in Europa*, it has repeatedly made clear that it understands Europe from its relationship of tension with other continents, with other religions, with other peoples and cultures.

The project of Amo's philosophy, however rudimentary it may have been, developed in the writings that have come down to us, certainly allows us to think the differences within a *République des Lettres* that is by no means necessarily limited to Europe, and to relate them to a horizon that ultimately delimits itself to the existence of the philosophies of the world. Amo's philosophy(s) had undoubtedly reached a very exciting point here in the *Tractatus*, but one that could also prove highly fraught with tension and danger for its author. For the essentially polylogical structure of his philosophizing, for thinking of difference that was in the making, there were contemporary limits.

Jena or Philosophizing at the Borders

The fact that Anton Wilhelm Amo, nevertheless, left Prussian Halle for a second time, despite all the esteem in which he was held, and went to the Universität Jena (University of Jena) for the first time, is certainly not due to a lack of qualifications in the field of research or teaching but is probably primarily due to the deteriorating environment in which Amo's precarious situation could become immediately dangerous for him. This time, however, the destination for his second departure from Halle was no longer Wittenberg, but the Universität Jena (University of Jena): And thus, beyond the previous changes of location, a movement that no longer knew a fixed point of reference, a fixed place of residence. The philosopher was not to return to Halle, but he was to return to Prussia. The last phase of Anton Wilhelm Amo's life in Germany, as we know it, had begun.

[72] Amo, Anton Wilhelm: *Traktat*, p. 118; in the original Latin: "*pro diversitate Gentium*" (*Tractatus*, p. 15).

[73] See on the development of this concept Ette, Ottmar: *ZwischenWeltenSchreiben. Literaturen ohne festen Wohnsitz (ÜberLebenswissen II)* (2005).

On June 27, 1739, Amo submitted his original[74] application for nostrification to the Universität Jena (University of Jena), a request signed "Anton Wilhelm Amo, an African, Master and Lecturer of Philosophy and Liberal Arts and Rights Candidate", which was quickly circulated among the faculty there with the request for a statement and already resulted in a positive decision on July 8, 1739. Thus, only a few days later, on July 17 of the same year, Amo was able to publicly post his lecture announcement, which was of course written in Latin, at the university. The programme of the announced event is extremely wide-ranging and basically represents central parts of the philosopher's scientific activities. That Amo also ventured into the field of cryptography cannot really be surprising in view of his experience in cryptographic writing.

At the same time, however, we should note that Amo posted the annotated announcement of the lecture at the university that was still regarded in Halle as "the place of 'Wolffian corruption' of the royal Prussian children".[75] One can assume that Amo expected, or at least hoped for, a more open intellectual climate in Jena, both for his concerns and himself. Amo's reference to his own financially precarious situation and the absence of any reference to the Dukes of Brunswick-Wolfenbüttel make it clear once again in his application to the Universität Jena (University of Jena) that his situation was no longer comparable to that of the young student and prospective Magister.

The status of servitude (at least slavery-like) no longer existed in a visible way. Professors Wideburg and Stolle, as advocates of the nostrification of Amo in Jena, pointed out, among other things, that Amo deserved the "commiseration" because he had been "abducted from another part of the world in his tender youth" and had been "converted to the Christian religion" by "heydenthum" (paganism).[76] A plea for the granting of a university reprieve? It almost seems so. For Amo had no parents and no relatives to whom he could turn but wanted "now not to beg", but to make an honest living based on his abilities.[77]

Thus, the deported enslaved boy seems to have become, at first and for a few probably quite happy years, an aspiring young scientist, but soon a penniless private lecturer left to his own devices, who—this may have become apparent to Amo in Jena at the latest—would also have little chance of a secure position within the university landscape and, in particular, the professoriate in Germany. The author of the *Tractatus*, whose perseverance impresses, again and again, signed the lecture announcement in Jena, written down by his hand, with his self-chosen name affix 'Afer' and his status as 'Magister legens'.[78]

---

[74] Reprint of the facsimile in Brentjes, Burchard: *Anton Wilhelm Amo. Der schwarze Philosoph in Halle*, p. 60.

[75] Ibid., p. 61.

[76] Quoted from ibid., p. 63.

[77] Ibid.

[78] Reprint of the facsimile in Brentjes, Burchard: *Anton Wilhelm Amo. Der schwarze Philosoph in Halle*, p. 66.

With the rise to power of Frederick II of Prussia, who recalled Christian Wolff, who had been banished by his father, to Halle, Prussia entered a period of intense intellectual impetus in the context of an Enlightenment movement that was to be understood not only in Europe but also worldwide, but there is no evidence that Amo could have attempted to return to Halle a third time or, for example, to transfer to the Viadrina in Frankfurt on the order—as is well known, Berlin did not have its university at the time.

However: There is very well the already at the beginning of these considerations mentioned hint of the usually informed Johann Friedrich Blumenbach that Amo had probably gone as court counsellor[79] to the Prussian court in Berlin. William E. Abraham speculated here—as Wolfram Suchier had done before him[80]—that the call to the court in Berlin and Potsdam might have come from the young Frederick II (who succeeded his father on the Prussian throne in 1740), since the young monarch, committed to the Enlightenment, had also brought Wolff back to Prussia from his banishment.[81]

A possible stay of Amo in Berlin or Potsdam could indeed also explain why there were no more signs of life of Amo from the university cities Halle, Wittenberg or Jena.[82] However, there is no evidence of Amo's activity at the court of the young Frederick (so far). Among the few signs of the philosopher's life, there is probably a later hint that Amo is said to have been mentioned for the last time in Jena in 1747 as a citizen living there.[83] This would not necessarily exclude a temporary stay in Berlin.

So much speaks for a 'return' of Amo to Prussia in search of a better position at court or elsewhere. Certainly, however, it would never have occurred to him to join the services of those 'Court Moors' who did their duty at the Prussian court. Even today, the so-called 'Mohrenrondell' in the park of Sanssouci, from which the cover motif is also taken, bears witness to the servile presence of enslaved blacks without slavery at the Prussian court. It is tempting to imagine the philosopher, educated in Prussia, in Berlin or at Sanssouci during that period that is obscure in Amo's life. But it will take further research to substantiate or definitively falsify such notions, which go back to the usually well-informed Blumenbach. It is striking that we, unfortunately, have no concrete evidence from contemporary witnesses or contemporaries of the habilitated scholar's presence in enlightened Prussian climes.

---

[79] See Blumenbach, Johann Friedrich: Von den Negern. Einige naturhistorische Bemerkungen bey Gelegenheit einer Schweizerreise, p. 9; Blumenbach's account, however, is not considered very credible by Mabe, Jacob Emmanuel: *Anton Wilhelm Amo interkulturell gelesen*, p. 34.

[80] See Suchier, Wolfram: Weiteres über den Mohren Amo. In: *Altsachsen. Zeitschrift des Altsachsenbundes für Heimatschutz und Heimatkunde* (Holzminden) 1–2 (January–February 1918), pp. 7–9, p. 9. However, contrary to Suchier's hopes, no clarification has yet been found in the Geheimen Staatsarchiv (Secret State Archives) in Berlin.

[81] See Abraham, William E.: Anton Wilhelm Amo, p. 197.

[82] See ibid., p. 197.

[83] See Brentjes, Burchard: *Anton Wilhelm Amo. Der schwarze Philosoph in Halle*, p. 66.

Amo must have realised early on that it would be difficult for him to improve the conditions for his work. As early as May 5, 1740, he wrote to his student and friend Gottfried Achenwall, the later famous founder of statistics in Germany, a sentence of Epictetus, whom he admired, in his family book: "Necessitati qui se accomodat sapit, estque rerum Divinarum conscius."[84] And he once more signed his entry "Antonius Guilielmus Amo Afer/Philos. et Art. Liberal. Magister Legens."[85] The stoic composure that Amo displayed here may make it clear, against the background of the certainly stoic character of his philosophy,[86] that he had long understood how narrow were the limits that the academic field had set for his philosophy, for his philosophizing. One could translate the dictum of Epictetus, the late representative of Stoic philosophy, as follows: 'He who knows how to adapt himself to necessity is wise and aware of divine things'.

The forties were certainly not a good time for Amo in Germany. His last great patron, Johann Peter von Ludewig, died in 1743, which certainly deprived him of many hopes of finding a secure position at a university in Germany. After Ludewig's death, Amo's situation probably worsened. In 1747, a nasty piece of invective was even written about him, which undoubtedly contained proto-racist attacks against him. At the center of this slanderous piece was the learned Moor Amo, who makes unspeakable proposals to a beautiful brunette. But the German girl laughs at the suitor for his physical attributes and sends the thus ridiculed one back home.[87] The lampoon gives us an idea of that atmosphere in which Anton Wilhelm Amo, who had by now move to Germany, lands. His brilliant erudition only served his envious and mockers to make fun of him even more.

The author of these slurs in rhymed form against the 'learned Moor' was a former fellow student of Anton Wilhelm Amo, the temporary professor of German language and eloquence Johann Ernst Philippi.[88] One should not attach too much weight to this machination. For it is more than questionable whether one may conclude from it a real incident of the rejection of Amo as a lover-to-be since mockeries in the good literary tradition of the mocking poem often resorted to a brusquely

[84] Cited in Suchier, Wolfram: A.W. Amo, S. 446. A second, long unknown entry in the family register from March 3, 1746 draws attention to this. Firla, Monika: *Ein Jenaer Stammbucheintrag des schwarzen Philosophen Anton Wilhelm Amo aus dem Jahr 1746*, S. 16 f. There Firla also speaks of a synthesis between Epictetus and Christianity striven for by Amo (ibid., p. 21).

[85] Cited in Suchier, Wolfram: A.W. Amo, p. 446.

[86] See Firla, Monika: *Ein Jenaer Stammbucheintrag des schwarzen Philosophen Anton Wilhelm Amo aus dem Jahr 1746*, S. 7 f.

[87] A reprint of this literarily bumbling lampoon can be found at http://www.theamoproject.org .

[88] Brentjes, Burchard: *Anton Wilhelm Amo. Der schwarze Philosoph in Halle*, p. 70; for more details see Abraham, William E.: Anton Wilhelm Amo, p. 197 f. The sources for the story told by Johannes Glötzner of a love that also reaches its climax with a common child and thus a descendant of Anton Wilhelm Amo in Germany are unclear: see Glötzner, Johannes: *Anton Wilhelm Amo. Ein Philosoph aus Afrika im Deutschland des 18. Jahrhunderts*, p. 17. Is this just another of the now quite numerous Amo legends? Not entirely innocent of numerous embellishments could be the fact that Johannes Glötzner is the author of probably the first novel dedicated to Anton Wilhelm Amo. The lampoon gave rise to the most diverse speculations.

rejected courtship of a man mostly physically disqualified (predominantly by age, stupidity or ugliness). The real evidence for the validity of such an incident, which certainly sounds seductive in its novelistic setting, does not exist—as far as I can see. But the existence of this piece testifies beyond all doubt to the growing hostility to which the philosopher was exposed in Germany.

This messy 'affair' provides a glimpse of the unmistakably racist prejudices and also many an attack to which Anton Wilhelm Amo may have been subjected in the course of his life in Europe and not least during his time at German universities. That the black German philosopher and 'doctissimus magister legens' was a public figure who unquestionably attracted enormous attention may certainly be attested to by Philippi's diatribes. But by the time these mocking verses and occasional sermons appeared, Amo, far superior in intellectual powers to his opponents, had—as we shall see—long since left the university town of Jena, Prussia, Germany and Europe behind him. And as the first phase of his life had ended abruptly with a voyage, so the last phase of this life's journey began again with the boarding of a ship. Anton Wilhelm Amo had made a decision. Was he, from his point of view, following Epictetus' dictum? In any case, he was philosophically facing a formidable challenge that would once again take him far beyond the boundaries of his time.

Africa or the Journey to an Unknown Origin

A family book entry by the hand of Amo dated March 3, 1746, discussed in detail by Monika Firla,[89] proves beyond doubt that Anton Wilhelm Amo was still in Germany at this time. However, he had perhaps already submitted an application or petition to the Dutch West India Company at this time, requesting free passage to the Gold Coast.[90] The powerful trading company that had once snatched him away to Europe was to transport him back to the shores of Africa. It was a journey to a land unknown to him.

This request was granted. And so, Amo might have left Germany in late 1746—Jacob Emmanuel Mabe, however, gives as date of departure from Rotterdam December 20, 1746[91]—or at the latest at the beginning of[92] the year 1747. Since Amo's ship reached the West African coast already in April 1747—Mabe again precisely dates it to April 7 of this year[93]—he probably embarked in Holland immediately after leaving Germany and reached Axim in present-day Ghana without any

[89] See Firla, Monika: *Ein Jenaer Stammbucheintrag des schwarzen Philosophen Anton Wilhelm Amo aus dem Jahr 1746*, pp. 15–22.

[90] Ibid., p. 25. See also Mabe, Jacob Emmanuel: *Anton Wilhelm Amo interkulturell gelesen*, p. 26.

[91] Ibid.

[92] Firla, Monika: *Ein Jenaer Stammbucheintrag des schwarzen Philosophen Anton wilhelm Amo aus dem Jahr 1746*, p. 25.

[93] See Mabe, Jacob Emmanuel: *Anton Wilhelm Amo interkulturell gelesen*, p. 26.

problems after safe navigation.[94] A *retour au pays natal*? *Back to Africa*? So, had Amo started his journey home?

The most important source for our knowledge about the last years of Amo's life in West Africa is still the report of the Swiss traveller and ship's doctor in Dutch service, David Henri Gallandat. The notes he recorded in 1752 appeared in the *Verhandelingen, uitgegeven door het Zeeuwsch Genootschap der Wetenschappen te Vlissingen* in 1782.[95] There is no doubt that the core of the news gathered there can only have come from the mouth of Amo, but the orality of this conversation was translated by Gallandat into written notes, which in turn were printed decades later in 'Low German', i.e. Dutch. However we may evaluate the individual biographies listed and judge the veracity of the references to father, sister, and brother Anton Wilhelm Amo, it becomes clear that they open up sufficient room for different speculations in each case.

However, there are also broad zones of consensus in the research devoted to Amo. What all interpreters of Anton Wilhelm Amo agree on—as far as I can see—is the fact, which seems almost 'natural', that the philosopher has returned or come home to his 'homeland'. The associated movement figure of a circle, which generates meaning as a hermeneutic movement pattern,[96] would thus be completed. This would create a life cycle that could begin with the forcibly enforced displacement and deportation as a slave from one's homeland and find its conclusion with the voluntary return to one's homeland that was undertaken willingly and intentionally. But can such an understanding of Amo's entire *curriculum vitae* appear to be sound? And was it really a homecoming?

Certainly, the attribution of the notion of a 'return to his fatherland' to the figure of Amo's movement in the publication of David Henri Gallandat's notes, in which

---

[94] For more on Amo's ship voyage and his 'return home' see Firla, Monika: *Ein Jenaer Stammbucheintrag des schwarzen Philosophen Anton Wilhelm Amo aus dem Jahr 1746*, pp. 26–31.

[95] Quoted from Brentjes, Burchard: *Anton Wilhelm Amo. Der schwarze Philosoph in Halle*, p. 68 f. Because of its great importance for Amo research, this passage is also quoted in the original: "Terwyl hy op deeze reis te Axim op de Goudkust in Africa was, ging hy den beroemden Heer Anthonius Guilielmus Amo Guinea Afer, Philosophiae Dr. et Artium Liberalium Magister bezoeken. Hy was een Neger, die ruim 30 Jaaren in Europa verkeerd had. Hy was in den Jaare 1707 in Amsterdam, en werd vereerd aan den Hertog van Brunswyk, Anthoni Ulrich, die hem aan zyn zoon Augustus Wilhelmus gaf. Deeze liet hem studeeren te Hall; en in Wittenberg, waar hy in den Jaare 1727 tot Doctor in de Philosophie en Meester in de Vrye konsten Gepromoveerd werd. Eenigen tyd daarna overleed zyn Meester: dit maakte hem zeer droefgeestig, en deed hem bessuiten naar zyn Vaderland te rug te keeren; hy leefde daer toen als een Heremiet, en had den naam van een Gelukzegger te zyn; hy sprak verscheiden taalen, Hebreeuws, Grieks, Latyn, Fransch, Hoog-en Nederduitsch; was zeer kundig in de Astrologie en Astronomie, en een groot Wysgeer; zynde toen omtrent 50 Jaaren oud. Zyn Vader en eene Zuster leefden noch, en woonden vier dagreizen landward in; hy had een Broeder, die Slaaf was in de Colonie van Suriname; naderhand is hy van Axim verhuist en gaan woonen in de Fortres der West-Ind. Comp. St. Sebastiaan, te Chama" (*Verhandelingen uitgegeven door het Zeeuwsch Genootschap der Wetenschappen te Vlissingen*. Middelburg: Pieter Gillissen 1782, pp. XIX–XX).

[96] On the significance of the different movement patterns and movement figures, see the opening chapter in Ette, Ottmar: *Literatur in Bewegung. Raum und Dynamik grenzüberschreitenden Schreibens in Europa und Amerika*. Weilerswist: Velbrück Wissenschaft 2001.

this momentous designation first appears in written form, can be justified by reference to the Itinerarium, to the simple movement in space. For it was to the vicinity of Axim that Amo was born on what was then the 'Gold Coast', and to the vicinity of Axim that he returned at the end of his life.

But even the appearance of the figure of the father, who is associated with the 'fatherland', which at first glance seems rather incidental, should draw our attention in this narrative to the fact that we are dealing here with a hermeneutic figure of movement that can be found in highly effective narrative patterns that have been influential in Western culture for centuries. The Christian parable of the prodigal son is only one of the countless variants that unmistakably resonate in this textual excerpt.

Moreover, recourse to the concept of the fatherland configures a narrative that has the strongest political and affective impact. For does it not seem 'natural' to return to one's fatherland? Even if such a conceptualization makes a mockery of the colonial conditions on the 'Gold Coast' at the time, its effect, at least on a European reading public that this text was intended to reach, is undeniable. The text attributed to Gallandat suggests on a variety of levels how 'natural' the return of the 'prodigal' son to 'his' 'fatherland' is. It is obvious that colonialist patterns of perception and representation are concealed behind this. The *Back to Africa* movement is not always meant to be anti-colonialist and emancipatory. It can very well correspond to colonialist interests.

Yet Anton Wilhelm Amo might very well have been the first among all interpreters who—when he formulated the request to the West India Company—would have conceived of his own life story long before Gallandat according to this model of a homecoming of the prodigal son, a return to his fatherland, or a journey to his childhood. For did not the return made in space (and only in space) ultimately mean a homecoming to a world he had once inhabited? Could not he thus close the cycle of his life and give his *curriculum vitae* a direction and thus a meaning—as is well known, the French *sens* stands for the direction as well as for the meaning?

Especially in the case of an author like Amo, who appealed to Christianity in his philosophical writings—certainly also in an academic field that was at times dominated by pietist power blocs—the idea of the return of the prodigal son in conjunction with other narratives could not but be effective. It is possible that these narrative models, along with the difficult living conditions in Germany, formed a central *movens* for this and spurred Amo's decision to face what was perhaps the greatest challenge of his life. Had not Amo himself pointed out the great importance of fiction in his reflections on the importance of ideas? Thus, his *Traktat says*: "The twin form of the possible idea is fictitious; it is a different conception of a thing from what it is in itself."[97] But the relation between a thing in itself[98] and its idea could indeed be thought of in Amo's sense as an equipollente, as an equally powerful one.

---

97 Amo, Anton Wilhelm: *Traktat*, p. 171; in the original Latin: "Gemina possibili idea ficta, quae: Repraesentatio rei aliter ac secundum se est" (p. 79).

98 On the meaning of the 'thing-in-itself' in Amo's philosophy and the history of philosophy in the eighteenth century, see Edeh, Yawovi Emmanuel: *Die Grundlagen der philosophischen Schriften*

One can only admire Amo for the courage to take on such a journey and thus such a risk. Was it a cultural shock that awaited him in Africa? Did he constitute a cultural shock to all who came into contact with him? Did the colonial authorities and slave traders of various stripes see him as a threat to their interests, or did he become an important interlocutor or mediator for some? Did Amo play a role there as a 'fortune teller' and was he able to become a person of respect for the members of different ethnic groups on the colonially enforced and dependent coast?

We do not know and should refrain from too far-reaching speculations. But it may be regarded as certain that Anton Wilhelm Amo repeatedly changed his place of residence not only in Germany but also in Africa. And it is probably one of the great contradictions and mysteries of his life that he spent the end of his existence—free or forced—in the fort of Chama and thus in a fortress that was integrated into the worldwide system of the slave trade under the rule of the West India Company. What we also know very well is that Amo's journey on board a ship of the West India Company was a journey to a country largely unknown to the traveller himself and certainly not a 'homecoming'—even if the journey was and is always stylised as such from the most diverse points of view and with the most diverse interests.

Writing-between-Worlds: Philosophizing Without a Fixed Abode

Let us ask ourselves the question that must be asked of every buried tradition: What remained of Anton Wilhelm Amo's intense philosophical and university activity in Prussia and Germany? The fame of Anton Wilhelm Amo Afer was by no means confined to the German-speaking world. In his famous comments on *De la littérature des nègres* from 1808, Abbé Henri Grégoire,[99] the Jacobin French philosopher and fighter against slavery and the slave trade, referred on no less than five pages to Anton Wilhelm Amo as well as to his origin from Guinea and his stay at the universities of Halle and Wittenberg. As in Zedler's *Grosser vollständiger Universal Lexicon aller Wissenschaften und Künste*, with which there are several similarities, attention is drawn to Amo's investigation of the legal situation of blacks in Europe right at the beginning, with explicit reference back to Blumenbach's remarks.

The philosopher's multilingualism is emphasized from the beginning: "Amo was well versed in astronomy and spoke Latin, Greek, Hebrew, French, Dutch and German."[100] Moreover, Grégoire resorted to the "hommage public" of the rector of

von Amo. In welchem Verhältnis steht Amo zu Christian Wolff, dass man ihn als einen ‚führnehmlichen Wolffianer' bezeichnen kann?, p. 70 f; as well as Schilling, Georg: "*Who was Anton Wilhelm Amo?*"—"*Remapping the Jungle*" *vs.* "*Vom faulen Holze lebend*"*?! An attempt against forgetting and* "*white*"*-washing, sad, racist* "*(a)e(sthe)tic(al)*" *texts of so called German* "*prime fathers*" et al., p. 15.

[99] See the reference in Damis, Christine: *Le philosophe connu pour sa peau noire: Anton Wilhelm Amo*, p. 119.

[100] Grégoire: Henri: *De la Littérature des nègres, ou Recherches sur leurs facultés intellectuelles, leurs qualités morales et leur littérature; suivies de Notices sur la vie et les ouvrages des Nègres*

the Universität Wittenberg (University of Wittenberg), who had rightly recalled "that Terence was also from Africa."[101] No doubt the French abbé knew very well that Terence had already given himself the epithet 'Afer'.

Even if Abbé Henri Grégoire clearly underestimates the duration of Amo's stay in Europe, the dimension of the geographical 'return' to the West African Gold Coast is mentioned and at the same time placed in the context of a 'deep melancholy', the French *mal du siècle* at the turn of the nineteenth century. Thus, for the first (but by no means the last) time, Amo is given the traits of a romantic hero who, at the age of fifty, had finally led 'la vie d'un solitaire'.[102] At the same time, Grégoire borrows the biographies of Amo's father and one of his sisters from the academician Gallandat he mentions, and Amo's brother was taken to Surinam as a slave.[103] With this, Grégoire has basically developed the essential biographical reference points that still structure any account of Amo today. Finally, the French *philosopher* emphasizes that his research has not enabled him to find out whether Amo published other works and when exactly he died.

By including Gallandat's travelogue, Grégoire succeeds for the first time in bringing to light a circular structure in the life of Anton Wilhelm Amo, a life that began in Africa and ended in Africa. A glance at the further history of Amo's reception can easily demonstrate how much this cycle was suited to be romanticized, idealized, or placed under the sign of a *Back to Africa* in subsequent times, a history that admittedly cannot be elaborated on here.

At the same time, this marked the beginning of a history of the most diverse instrumentalizations of Amo, which continues unabated to this day, and has even gained considerable momentum. Even Burchard Brentjes, who for his part did not shy away from reclaiming Anton Wilhelm Amo for the socialist policy of friendship between peoples of the German Democratic Republic in the times of the South African apartheid regime,[104] drew attention at the end of his monograph to an important line of reception in the twentieth century, since it had remained for the Reverend Attoh Ahuma and thus "an African" to "rediscover Amo at the beginning of the

qui se sont distingués dans les Sciences, les Lettres et les Arts. Paris: chez Maradan Libraire 1808, p. 198.

[101] Ibid., p. 199.

[102] Ibid. See here also the entry in the *Nouvelle Biographie Universelle* of 1852, to which attention was first drawn Struck, Bernhard: Nochmals A.W. Amo. In: *Akademische Rundschau* (Leipzig) V (1916–1917), pp. 54–56.

[103] Grégoire, Henri: *De la Littérature des Nègres*, p. 201.

[104] This applies to the opening and closing sections of his monograph *Anton Wilhelm Amo. Der schwarze Philosoph in Halle* as well as in particular for Brentjes, Burchard: Ein Afrikaner in Halle vor 250 Jahren? In Brentjes, Burchard (ed.): *Der Beitrag der Völker Afrikas zur Weltkultur. Materialien einer wissenschaftlichen Arbeitstagung zu Ehren des Philosophen Anton Wilhelm Amo (1727–1747 in Halle, Wittenberg und Jena).* Halle (Saale): Martin-Luther-Universität Halle-Wittenberg 1977, pp. 3–15. This in no way calls into question Brentje's great services to Amo research. Massive criticism of the instrumentalization of Amo for the politics of the GDR was expressed by Damis, Christine: *Le philosophe connu pour sa peau noire: Anton Wilhelm Amo*, pp. 121–123.

twentieth century" with his volume *Memoirs of West African celebrities*[105] published in Liverpool in 1905.[106] After the Second World War, this early account would have been followed up by "the forces of progress in the struggle against reaction".[107] Thus, beginning in 1946, the first "civil rights activists in the United States" to "draw on Amo"[108] would have been Beatrice Jackson Fleming and Marion Jackson Pryde with their book *Distinguished Negroes Abroad*.[109] In the same year, the great Kwame Nkrumah had also taken an interest in the "African philosopher"[110] and planned to write his book.

Consequently, in a sign of the solidarity of socialist anti-colonialists, Brentjes then also let the eminent African socialist have the last word by reprinting, instead of a concluding word, a letter from Kwame Nkrumah to him from Accra, dated May 14, 1964, concerning the great interest of the politician—"one of the leaders of liberating Africa"[111]—in Anton Wilhelm Amo. The letter says of Amo's time in Germany:

> He was persecuted for his philosophical views and hounded from one university to another by the German Pietists. Finally, in 1747, he was the victim of a racist attack in Jena and decided to return to Ghana.[112]

After his inclusion in the grand historical design of *De la littérature des nègres* and his consequent integration into the abolitionist project of the enterprising Enlightenment philosopher Henri Grégoire, Amo had now arrived in the second half of the twentieth century and thus in the debates surrounding the decolonization of Africa. Amo's image had changed once again: Persecuted for his philosophical views and a victim of racist attacks, the African had returned to his African homeland. Anton Wilhelm Amo's life was ideally suited not only to trace the stages of a non-poetic but philosophical *Cahier d'un retour au pays natal* in the spirit of Aimé Césaire's great creation, which appeared in the first version in 1939,[113] but also to illustrate it in a historically exemplary way. But what remained, what remains of Anton Wilhelm Amo and his life and creative period in Prussia? And what sense does it make to leave it at the erasure of his name, his work and his person in Germany?

---

[105] Ahuma, S. R. B. Attoh: *Memoirs of West African celebrities: Europe, & c., 1700–1850.* With special references to the Gold Coast. Liverpool: D. Marples 1905.

[106] Brentjes, Burchard: *Anton Wilhelm Amo. Der schwarze Philosoph in Halle*, p. 73.

[107] Ibid.

[108] Brentjes, Burchard: *Anton Wilhelm Amo. Der schwarze Philosoph in Halle*, p. 73.

[109] See Fleming, Beatrice Jackson/Pryde, Marion Jackson: *Distinguished Negroes Abroad.* Toronto: Associated Publishers 1946, pp. 148–152.

[110] Ibid.

[111] Ibid., p. 75.

[112] Nkrumah, Kwame: Letter of May 14, 1964, reprinted in Brentjes, Burchard: *Anton Wilhelm Amo. Der schwarze Philosoph in Halle*, p. 77. See also Brentjes, Burchard: *Consciencism: Philosophy and ideology for De-Colonization.* New York/London: Monthly Review Press 1970.

[113] See Césaire, Aimé: Cahier d'un retour au pays natal. In: *Volontés* (Paris) 20 (1939).

Certainly, it is possible—and has been practiced many times—to *reterritorialize* Anton Wilhelm Amo starting from his birth (and not from the major part of his life) in such a way that he alone becomes an African in fulfillment of his self-chosen epithet. But it seems a highly dubious procedure to tie a human being who has been *reterritorialized in* this way back, as it were genetically and biologically, to an 'African mental world', however, constructed,[114] and thus to reduce him to a single 'identity' and affiliation (namely that of his origin). If one follows Kwasi Wiredu, Amo's 'homecoming' was about cultural survival, since Amo was an African and a philosopher, and thus an African philosopher.[115] The talk of unconscious African traces and a "racial identity"[116] aims at a "return" consciously linked to the "homecoming", also and especially as a philosopher, which struggles to stylize the author of *De iure Maurorum in Europe* as an exclusively African thinker.

On the other hand, it could be noted that this movement of inclusion is matched by a movement of exclusion in a not coincidentally complementary way. For it would be easy to prove that Anton Wilhelm Amo Afer's writings and scholarly activities have been largely erased from the history of philosophy: Even the biographical entries that had appeared during Amo's lifetime, for example, in Zedler's *Großes vollständiges Universal Lexicon aller Wissenschaften und Künste* disappeared, as did his name, and even the memory of his name. His exclusion from the history of Prussia and Germany was complete and radical.

One can only agree with the actual 'rediscoverer' of Amo for the at least German-speaking area, the librarian Wolfram Suchier when he emphasizes that the study of Amo will yield "further building blocks for the history of German scholars".[117] This preoccupation is well-founded, however, since Amo is "memorable" because of "the sensation he caused in Germany in his time, and because of his writings for the history of universities, scholars and culture."[118] The inclusion carried out with regard to Africa and the exclusion carried out very extensively with regard to the history of German and European science, thought, and philosophy is complementarily interwoven processes that need to be broken down in their mutual support. For they are always based on simple territorializations and exclusive attributions as well as exclusions, which in many cases knowingly and willfully seek to reduce the complexity of the life and work of Anton Wilhelm Amo.

The systematic and still effective erasure or exclusion of Amo from the history of German and European philosophy corresponds on a structural level to the attempt to stylize Amo as a thinker of a purely African tradition and identity. But we should recognize from his thinking, from his writing, that Anton Wilhelm Amo, as a European thinker and as a representative of the German early Enlightenment, was at the same time an African philosopher, undermining any attempt to conceive of him as either 'merely' one or 'merely' the other. Fortunately, the times when he was neither one nor the other are definitely over.

---

[114] See et al. Mabe, Jacob Emmanuel: *Anton Wilhelm Amo interkulturell gelesen*, p. 100.

[115] See Wiredu, Kwasi: *Amo's Critique of Descartes' Philosophy of Mind*, p. 205.

[116] Ibid.

[117] Suchier, Wolfram: *Weiteres über den Mohren Amo*, p. 7.

[118] Ibid.

Yet Amo's entire oeuvre cannot simply be reduced to the common denominator of a purely additive both/and. Could this not be related to the fact that intercultural philosophy has so far shown itself so little receptive to his thinking and has also largely excluded him?[119] For Anton Wilhelm Amo is far more than the addition of both inscriptions and traditions, but rather the bold and at the same time always precarious attempt to be not the addition, but the *movement* between these strands of tradition, and to participate explicitly as an African in the German early Enlightenment—and without allowing himself to be reduced to it alone. It is precisely here that Amo's significance lies in the history of ideas as well as the history *tout court* of Prussia and Germany.

Amo's philosophy is, in the sense of his *Tractatus de arte sobrie et accurate philosophandi*, an art of *philosophizing*, of thinking and writing out of the movement, of a transareal dynamic that cannot be stopped. Amo not infrequently cryptographically made use of this art—a fact that necessitates intensified philosophical as well as philological attempts to decipher and deal with this author. For his philosophy is philosophizing without a fixed abode, which cannot be reduced to a single perspectivity, to a single affiliation, to a single standpoint.

But precisely therein lies the promise of this philosophizing for the future. We have seen under what difficult conditions of inhuman slavery and exclusion, princely funding of an experimental object and university trench warfare in the academic field, the art of this sober and calm philosophizing had to develop and sought to develop. But the perseverance of a resistance that always took the difficulties as an occasion for new efforts gave this art its consciousness, indeed its self-awareness.

The transareal vectorization of knowledge of life, which tested itself as knowledge of survival precisely in the field of thought, strongly determined Amo's philosophizing, insofar as the crossing of different languages, diverse cultures and often contradictory traditions gave rise to a dynamic polylogic, which thoughtfully but also emphatically placed theology as well as philosophy in the plural in Amo's thinking. At the same time, he experienced at German universities how his way of Writing-between-Worlds was not able to protect him from exclusion. Leaving Germany was undoubtedly preceded by a collapse of livable conviviality.

Philosophizing, as practiced by Antonius Guilielmus Amo Afer in Latin, has no fixed place, no secure coordinate system, and certainly no 'home', which never existed in Amo's life anyway. Unlike other languages of philosophy, Latin in the eighteenth century was not committed to any place or territory. Anton Wilhelm Amo's philosophizing without a fixed abode multiplies the places of his own life, thinking and writing through its constant movements, and is oriented less towards a retrospective history of philosophy than towards a prospective shifting of horizons, thinking into the future and creating new futures. Thus, this art of philosophizing forms in the past a piece of the future. A future that needs to be uncovered from Amo's buried past future for Germany and Europe that will have shed all colonialism and understand Anton Wilhelm Amo as a historical figure who is also part of our history and asks us burning questions.

[119] See Mabe, Jacob Emmanuel: *Anton Wilhelm Amo interkulturell gelesen*, p. 103.

Frederick the Great and Cornelius de Pauw: Prussia, Mexico and the (New) World

In the Universe of Texts

Cornelius de Pauw may undoubtedly be considered one of the most famous representatives of the European Enlightenment among his contemporaries worldwide. His writings, which often caused a sensation on a global scale, were read in the context of the 'dispute about the New World', the history of which Antonello Gerbi has reappraised in a study that has long since become a classic, in a *République des Lettres* that quite naturally spanned national boundaries.[1] In short: Without de Pauw, a decisive phase of this dispute, the Berlin debate about the New World, would be inconceivable.

Yet the multitude of polemics that de Pauw triggered, especially with his *Recherches philosophiques sur les Américains*, seem long forgotten today, even if one still occasionally finds his name in fleeting footnotes or casual asides. A glance at the international research landscape reveals that there are hardly any recent publications on the works of the philosopher, who was born in Amsterdam in 1739 and died in Xanten in 1799, let alone more extended monographs that would examine his entire philosophical oeuvre.[2] It has become quiet, even suspiciously quiet, about the man who once heated minds and sparked one of the undoubtedly decisive disputes of the century of Enlightenment. Do his texts no longer have anything to say to us today?

The questions with which de Pauw was concerned over long periods of his work are certainly still of the greatest relevance today, especially in their cultural-philosophical thrust, if one wants to understand the history of relations between Europe and the non-European world with historical depth. For the philosopher, who was feared for his sharp-tonguedness, is a thinker of globality, and his

[1] See also Gerbi, Antonello: : *La Disputa del Nuovo Mondo. Storia di una Polemica: 1750–1900.* New edition edited by Sandro Gerbi. Milan/Naples: Riccardo Ricciardi Editore 1983.

[2] The work of de Pauw is mentioned at least in Cañizares-Esguerra, Jorge: *How to Write the History of the New World. Histories, Epistemologies, and Identities, in the Eighteenth-Century Atlantic World.* Stanford: Stanford University Press 2001, pp. 26–28.

O. Ette, *Mobile Prussia*, https://doi.org/10.1007/978-3-476-05839-3_2

work is a response to that second phase of accelerated globalization,[3] which encompasses the second half of the eighteenth century until well beyond the turn of the century, a period which, incidentally, has also been reflected within the German-speaking world since the middle of the century in a marked increase in the number of translations of Spanish historical works on the first phase of globalization.[4]

Unity and the division of the entire planet are problems that pervade de Pauw's entire work on both the Old and the New World. In the process, the author of the *Recherches philosophiques sur les Américains* also understands this 'New' World as a geologically and geomorphologically younger, newer world, entirely in the sense of Buffon's natural history:

> The almost infinite number of lakes and swamps with which the West Indies are covered, were not formed solely by the extravasated river waters, nor by the fogs attracted by the mountains and forests: These lakes appear to be deposits of water which had not yet been able to flow out of the places formerly drowned by a violent tremor, imparted to the whole machine of the terraced globe: The numerous volcanoes of the Cordilleras & rocks of Mexico, the tremors which never cease in the one or other branch of the Andes, prove that the earth is not yet at rest there in our days.[5]

The entire planet thus forms a unity, that 'machine of the globe', which, of course, from the point of view of natural history, is at the same time in an earth-historical non-simultaneity. For while in the Old World a certain calming ("repos") had long since come about, it is seething and shaking in the mountains of the New World, where the land masses would have risen from the water only much later and still formed wide aquatic inland spaces. The differences between the worlds are consequently immediately perceptible to the senses. 'Old' is directly opposed to the 'New'. The machine of the earth's body is indeed one, but it splits into two hemispheres, which in their opposition persist to our days and in their violence become sensually perceptible to the human body as a result of the numerous earth tremors.

The distinction and separation of two hemispheres forms the fundamental discursive setting of the *Recherches philosophiques*. It is initially rational, but not

---

[3] See also for the first time Ette, Ottmar: *Weltbewußtsein. Alexander von Humboldt und das unvollendete Projekt einer anderen Moderne.* Weilerswist: Velbrück Wissenschaft 2002, pp. 25–27; a comprehensive overview of the different phases of accelerated globalization can be found in Ette, Ottmar: *TransArea: A Literary History of Globalization.* Translated by Mark W. Person. Berlin/Boston: De Gruyter Mouton 2016.

[4] See also Birkenmaier, Anke: *Versions of Montezuma. Lateinamerika in der historischen Imagination des 19. Jahrhunderts. With the complete manuscript of Oswald Spengler's 'Montezuma: ein Trauerspiel' (1897).* Berlin/New York: de Gruyter 2011, p. 14.

[5] Pauw, Cornelius de: *Recherches philosophiques sur les Américains, ou Mémoires intéressants pour servir à l'Histoire de l'Espèce humaine.* 2 vols. Berlin: Chez Georges Jacques Decker, Imp. du Roi 1768–1769, here vol. 1, p. 102. Der Band erschien unter dem leicht auflösbaren Pseudonym „Mr. de P***". Eine deutsche Ausgabe folgte im Übrigen rasch: *Philosophische Untersuchungen über die Amerikaner, oder wichtige Beyträge zur Geschichte des menschlichen Geschlechts.* Übersetzt von Carl Gottlieb Lessing. 2 vols. Berlin: Decker und Winter 1769. Die französischsprachigen Originalausgaben verfügen über zahlreiche orthographische Nachlässigkeiten insbesondere bei der Akzentsetzung, die im vorliegenden Text nicht gesondert signalisiert werden

relational in nature. It gives human history a meaning, but—as we shall see—quickly uncouples it from the senses. De Pauw's *Recherches* relies on the power of meaning, not on the power of the senses and of what can be experienced sensually by the traveller as well as by the audience.

The argumentation scheme that becomes clear in the above quotation is not only followed by the natural-historical, but also by the cultural-historical considerations of the *Philosophical Investigations* on those Americans (*Américains*), which of course do not designate—contrary to our current confusion of language—the inhabitants of the USA, but the indigenous population of the American continent. Right at the beginning of his 'Discours Préliminaire' to the first volume of his extensive writing, published in Berlin in 1768 under the name of a "Mr. de P***", de Pauw made it clear which historical event was for him the one that, from his point of view, had most lastingly shaped the history of the planet and, as it were, had most strongly brought forth the natural-historical difference between the Old and the New World:

> There is no event more memorable among men than the Discovery of America. In going back from the present to the most remote times, there is no event that can be compared to that one; and it is undoubtedly a great and terrible spectacle to see a half of this globe, so disgraced by nature, that everything there was either degenerate or monstrous.
>
> Which physicist of antiquity ever suspected that the same planet had two such different hemispheres, one of which would be defeated, subjugated & as if swallowed up by the other, as soon as it was known, after a lapse of centuries that are lost in the night & abyss of time?
>
> This astonishing revolution, which changed the face of the earth and the fortunes of nations, was absolutely momentary, because, by an almost inconceivable fate, there was no balance between attack and defense. All the strength and all the injustice were on the side of the Europeans: The Americans had only weakness: They were therefore to be exterminated and exterminated in an instant.[6]

Cornelius or Corneille de Pauw,[7] not for nothing educated at Jesuit colleges in Liège and Cologne and enrolled at the University of Göttingen best equipped for non-European questions, now developed a contradiction in the sharply contoured twists and turns of his French-language work, as always with him, which now transferred the opposition of 'old' and 'new' world to that of 'Europeans' and 'Americans'. They are inhabitants of *a* world that could hardly be more sharply divided into two.

The mechanisms of inclusion and exclusion discursively set in motion in this opening passage thus generate a structural ambivalence, which, however, is immediately taken back into a radical divorce as a result of the historical processes evoked here. In terms of the philosophy of history, it determines the entire layout of de Pauw's writing, which stands under the sign of a great clash that has basically still not found an end. The European powers meet the American ones—and truly not in

---

[6] Ibid., vol. I, p. a2v f.

[7] Vgl. hierzu Church, Henry Ward: Corneille de Pauw, and the controversy over his ‚Recherches philosophiques sur les Américains'. In: *PMLA* (New York) LI, 1 (March 1936), p. 180 f; sowie Beyerhaus, Gisbert: Abbé de Pauw und Friedrich der Große, eine Abrechnung mit Voltaire. In: *Historische Zeitschrift* (München/Berlin) 134 (1926), pp. 465–493.

a way that was later sought to be glossed over as a 'meeting of civilizations'. From de Pauw's perspective, the outcome of this clash was virtually predetermined by nature. This, however, gives the immense power in its asymmetry its historical-philosophical meaning.

Culture is consequently transformed into nature in this argumentation. And even more: The world of America is (re)transformed into a realm of nature, while Europe appears under the sign of a successfully advancing culture. From this, however, the New World is momentously excluded. What is contrarily conceived in terms of natural history manifests itself with almost deterministic violence on the level of cultural history. Thus the culturally generated power of the Europeans is statically and, as it were, absolutely opposed to the 'natural' powerlessness of the Americans. We are dealing here with that (deliberate) confusion of biology and culture that Claude Lévi-Strauss had so strongly emphasized in his discussion of 'race' and 'history':

> But the original sin of anthropology consists in the confusion between the purely biological notion of race (assuming, moreover, that, even on this limited terrain, this notion can claim objectivity, which modern genetics contests) and the sociological and psychological productions of human cultures.[8]

All life of nature as well as of culture is placed in the above text excerpt of de Pauw under the sign of this divorce, this construction as opposition, not forgetting, of course, that the two so different hemispheres still form the unity of a planet traversed by a deep separation. Thus Cornelius de Pauw already added to the *incipit* of his certainly most discussed and controversial work those fundamental isotopias which pervade his *Recherches philosophiques* in their entirety: A deliberate literary-rhetorical work, which precisely through its often polemical punch line—to which de Pauw owed his rapid fame as well as probably his no less rapid oblivion—did not fail to have its effect on the contemporary international readership. A greater contrast than that between the 'old' and the 'new' world is presented as hardly conceivable. The text unfolds the power of a single sense, a single direction of movement, whose vectoricity is clearly established and evident to the philosopher.

The long-term impact of de Pauw's analysis—even if the name of its author has disappeared from many works and compendia of the nineteenth century—should not be underestimated, nor should the variety of philosophical and political consequences it produced. Even Hegel's world-historical sketch, with its well-known thesis of the inferiority of America, its natural features, its animals and its indigenous population, shows the unmistakable traces of a way of thinking that Cornelius de Pauw had certainly taken to the argumentative extreme with his *Recherches philosophiques*.

In Hegel's *Lectures on the philosophy of world history*—and thus in the dominant philosophical tradition of Prussia and the German-speaking world—these ideas are still highly alive:

---

[8]Lévi-Strauss, Claude: *Race et histoire*. Suivi de L'œuvre de Claude Lévi-Strauss par Jean Pouillon. Paris: Denoël 1984, p. 10.

> The world is divided into the old and the new, and the name of the new has come from the fact that America and Australia became known to us only lately. But these parts of the world are not only relatively new, but new altogether, in view of their whole physical and spiritual constitution. [...] Of America and its culture, especially in Mexico and Peru, we have news, but only that it was an entirely natural one, which had to perish as soon as the spirit approached it. Physically and spiritually impotent, America has always shown herself and still shows herself so. [...] The inferiority of these individuals in every respect, even with regard to size, is apparent in everything; only the very southern tribes in Patagonia are stronger natures, but still entirely in the natural state of crudeness and savagery.[9]

Here, too, the culture of the Others is negated by being transformed back into nature and thus historically eliminated. It is enough that the (European) mind approaches it in order to make all this visible in its irrelevance for human history and to bring it to collapse. A study of one's own could undoubtedly substantiate the fundamental way in which the method, certainly not invented by de Pauw, but radiating very widely from his successful writings, shaped precisely the reflection on the relationship between the European and non-European worlds in the nineteenth century, indeed well into the twentieth century. But how can this method be described in more detail? And what consequences does it have?

With regard to Cornelius de Pauw's way of thinking and proceeding, it should first be noted that the other part of the world does not have its say, neither in the *incipit* nor in the further course of the *Recherches* itself. How should he do so? For, if one follows de Pauw, he lacks words. Thus it says in his influential contribution to the *Supplément* of the *Encyclopédie* with regard to the speechlessness that he attributes to the inhabitants of America:

> [...] the poverty of their language, whose dictionary could be written in one page, prevents them from explaining themselves. It is known that the Peruvians themselves, though united in a kind of political society, had not yet invented terms to express metaphysical beings, nor the moral qualities which should most distinguish man from beast, such as justice, gratitude, mercy. These qualities were among those things which had no name: virtue itself had no name in this country, about which so many exaggerations have been made. Now, among the little itinerant peoples, the scarcity of words is even greater, to the extent that any kind of explanation on matters of morals and metaphysics is impossible there.[10]

In these well-calculated turns of phrase, language—and the disposal of language—is brought into a fundamental relation to a not least political incapacitation, whereby the relation between linguistic ability and physicality is structured in the context of the entire passage. This is in a way similar to the relation between natural-historical and cultural-historical inferiority examined at the beginning, which—as we have already seen—disqualifies everything American as monstrous

[9] Hegel, Georg Wilhelm Friedrich: *Lectures on the Philosophy of World History.* Vol. XII. On the basis of the retained manuscript material, newly edited by Georg Lasson. Leipzig: Felix Meiner 21923, p. 107 f.

[10] Pauw, Cornelius de: America. In: : *Supplément à L'Encyclopédie ou Dictionnaires raisonné des Sciences, des Arts et des Métiers. Par une Société de Gens de Lettres. Put in order and published by M****. Volume 1. Amsterdam: Chez M.M. Rey, libraire 1776, p. 352.

or degenerate. The Dutch philosopher thus almost necessarily takes the place of those who have no words, no abstract concepts, and thus no power (and ability in any sense). He becomes the word-powerful, the word-powerful par excellence, and consequently speaks in place of those who allegedly do not have a sufficiently differentiated language at their disposal.

As numerous as the grammars and dictionaries of the most diverse American languages, mostly compiled by missionaries, had long been at that time: Cornelius de Pauw insists on the American "disette des mots" ("shortage of words") which gives him the power to substitute his own words for those of others. This profoundly sense-forming and sense-giving process can hardly be overestimated in its significance: The ostensible language specialist and philologist also in the further course speaks only *about*, but never *with the* (words of) the indigenous population, which—by the way—of course has not only *one*, but an extraordinarily high number of the most diverse languages on the American continent.

The *incipit* of de Pauw's *Recherches* already shows that the European is only able to understand itself if it places itself in a worldwide comparison (admittedly: According to its own rules and entirely in the sense of its own power). The outcome of this comparison is predetermined, a reciprocity between the two sides of the Atlantic is not foreseen. From now on, the lexeme *européen* fulfills the discursive task in the *recherches* of distinguishing one's own—certainly in its internal European diversity—from a non-European other, without the two coexisting quasi contactlessly.

For they are united by what tragically divides them: The history of an asymmetry of power, which by no means concerns only the evident inequality of their arsenals. But Europe, after the first phase of accelerated globalization, whose early historical climax was the so-called discovery of America by Columbus, can no longer be conceived without non-Europe. This is the point of departure from which the Amsterdam-born Abbé leads his readers into his New World. This is the sense that de Pauw tries with all his might to convey through his investigations—a sense that largely eludes the perception of the senses and entrusts itself solely to the universe of the texts.

The Power Over Meaning or the Berlin Debate About the New World

The *Recherches philosophiques sur les Américains* form a work of words that does not refer to any empirical object sensually experienced by de Pauw, but exclusively to other works and words, to other texts. Thus this writing stands in for what might be called a textual science in the weak sense. Neither did the Dutch abbé ever see or travel to the New World, nor did he—like Guillaume-Thomas Raynal, for instance, after him—maintain a wide-ranging network of correspondents, even overseas, and exchange information transatlantically. Cornelius de Pauw's method, it could be said, was purely text-based: In this sense it is philological, aiming at the sense and largely excluding the senses textually.

With good reason, therefore, one could characterize the *Recherches philosophiques sur les Américains* as a never-ending journey through the world of texts, but not through the world of sensual experience. Without any empirical basis, without any journey to his subjects, and without direct knowledge of the objects he described, discussed, and evaluated (in) the New World, de Pauw consulted historical chronicles as well as contemporary travelogues, board books accessible to him as well as manuscripts of commercial travelers, historiographical works of the sixteenth century along with philosophical treatises of the eighteenth century. In these movements between texts, there emerges not only a certain autonomy of the textual universes he traverses, but perhaps even more a textually generated autonomy and self-logic of a *philosopher* who makes his judgments from an elevated philosophical observer position and purports to speak in the name of a universal reason. This undoubtedly accounted for the fascination and also associated effectiveness of his work, which was always interested in clear meaning. It is based less on a mastery of objects than on a mastery of texts. De Pauw's philology could very well be understood as power, indeed as violence over texts.

Thus, de Pauw's own understanding of American history-as he noted in his aforementioned article for the Supplément volume of the Encyclopédie is grounded in a critical approach (if only anchored in the critic himself) to historical source texts, and thus in a specific quality of his own reading:

> When one wishes to have an idea of the state in which the new world was at the time of its discovery, one must study the reports, and constantly employ a judicious and severe criticism to discard the falsehoods and prodigies with which they abound: The compilers who have no spirit, pile up everything they find in the travelers' journals, & finally, make disgusting novels, which have only multiplied too much in our days; because it is easier to write without thinking, than to write while thinking.[11]

Only the critical reading of what has been written before makes possible a writing that reflects this observation: Reading and writing are the fundamental actions at the heart of de Pauw's textual universe. The norms for judging what counts as credible or must be excluded as mendacious in this world of texts can be defined solely from an enlightened Europe, indeed from Prussia. Americans are objects, but not subjects, of a non-reciprocal discourse that powerfully cuts them off and ascribes to their existence in the world-at-large a clearly delineated meaning as objects and as acted upon in history. Meaning is generated solely from Prussia, from Europe.

So what image of the (indigenous) American does de Pauw unfold on the basis of what he sees as text-critical work? In any case, it is a stereotypical image that has nothing to do with that of the *bon sauvage*:

> He is properly neither virtuous nor wicked: What motive would he have to be so? The timidity of his soul, the weakness of his mind, the necessity of procuring his subsistence in the midst of penury, the influence of his superstition, and the influences of the climate lead him astray, and lead him very far astray; but he does not realize this. His happiness lies in not thinking, in remaining in perfect inaction, in sleeping a great deal, in not caring about

[11] Pauw, Cornelius de: *Amérique (America)*, p. 353.

> anything when his hunger is appeased, and in caring only about the means of finding food when his appetite torments him. He would not build a hut, if the cold and the inclemency of the air did not force him to do so: he would not leave the hut, if he were not driven out of it by need: His reason does not grow old: He remains a child until death, foresees nothing, perfects nothing, and allows nature to degenerate in his eyes, under his hands, without ever encouraging it and without drawing it out of its slumber. Basically lazy by nature, he is vindictive through weakness, and atrocious in his vengeance, because he is himself insensitive: Having nothing to lose but his life, he regards all his enemies as his murderers.[12]

The *insensibilité* mentioned here stands—as will be shown in a moment—in a direct connection with a dullness of the senses, as de Pauw not only assumes but 'establishes' for the vast majority of non-European mankind. The reduction of the most diverse American cultures to the static image of 'the' American corresponds to the reduction of this human being to a quasi-animal existence, which is characterized by no development, no dynamics and no perfection. Culture is recoded into nature.

Thus 'the' original inhabitant of America, caught in an insurmountable immaturity, is *de facto* excluded from the history of mankind, to which he seems to contribute nothing, even more: He is expatriated from the human race, an exclusion that is even more radically developed in the second volume, also published in Berlin in 1769. Thus we read there of the human species—and it is not for nothing that the *Recherches philosophique* de Pauws, according to the subtitle, are about nothing less than the *Histoire de l'Espèce humaine*—in general:

> The true country where his species has always succeeded and prospered is the temperate northern zone of our hemisphere: It is the seat of his power, his greatness, and his glory. As we advance northward, his senses become numb and dull: The more his fibres and nerves gain in solidity and strength, by the action of the cold which tightens them; and the more his organs lose their finesse; the flame of genius seems to be extinguished in bodies that are too robust, where all the vital spirits are occupied in moving the springs of the structure and of the animal economy. [...] Under the Equator his complexion becomes tanned and blackened; the features of his disfigured physiognomy are revolting in their harshness: The fire of the climate shortens the term of his days, and by increasing the ardour of his passions, it narrows the sphere of his soul: He ceases to be able to govern himself, and does not emerge from childhood. In a word, he becomes a Negro, and this Negro becomes the slave of slaves.
>
> If we except the inhabitants of Europe, if we except four or five peoples of Asia, and a few small parts of Africa, the rest of the human race is composed only of individuals who resemble men less than wild animals: Nevertheless they occupy seven or eight times more space on the globe than all the civilized nations together, and almost never expatriate themselves. If Africans had not been transported to America in spite of themselves, they would never have gone there: The Hottentos do not travel any more than the Orangs [...].[13]

In this passage it is not without an unintentional humour that Cornelius de Pauw, who himself did not travel to the areas he dealt with, ascribes such a high, even decisive importance to travel for the development of cultural abilities and perfections. At the same time, it becomes clear that the majority of people are only

---

[12] Pauw, Cornelius de: *Recherches philosophiques sur les Américains*, vol. 1, p. 123.

[13] Pauw, Cornelius de: *Recherches philosophiques sur les Américains*, a.a.O., vol. II, p. 68 f.

included in the human race in order to be able to exclude them all the better afterwards and to place Europeans at the top of a humanity defined in this way. From the inferiority and 'eternal childhood' thus discursively produced, however, the right can very well be derived to assign them to a well-deserved slavery, which must necessarily be under the wise leadership of European humanity.

In this passage, as in others that follow, Cornelius de Pauw thus justifies European expansion at the expense of those tribes and hordes that occupy seven to eight times more space on the globe than all civilized peoples ("nations policées") combined.

The humanity that is thus ultimately divided into two is, however, just as indissolubly chained to one another as the divided world is, for better or worse, related to and interwoven with one another. Cornelius de Pauw thus unfolds the thinking of a globality that is irrevocably marked by a fundamental opposition and translates this opposition—on the scientific basis of Buffon's *Histoire naturelle*—into terms of complete European superiority and almost animalistic American inferiority.

This absolute European superiority is, however, subjected to fundamental criticism on the moral level in view of the first phase of accelerated globalization. Against the background of the colonial expansion of the fifteenth and sixteenth centuries, the author of the *Recherches* at the same time warns urgently of the consequences of that second wave of expansion spreading from Europe across the globe, as whose critical contemporary and spectator the Dutch abbé portrays himself. Europe was about to take possession of the "Terres Australes" (Southern territories) in terms of power politics and science under the guidance of "Politiques" (policies) and under the applause of some "Philosophes" (philosophers), without considering how much misfortune would necessarily be brought upon the peoples living there.[14] For the clarification of some disputed geographical questions or the verifiable execution of temperature measurements, one would be ready in this enlightened century at any time and without greater hesitation to accept "la destruction d'une partie du globe" (the destruction of a part of the globe).[15] In this way, Europe's science, in its independent logic, appears as both a source of impulses and a means of European expansionist policy. Science is recognized as part of (the problem of) European expansion itself.

But such a development, which is indisputably accelerating at the present time, must be opposed: "Mettons des bornes à la fureur de tout envahir, pour tout connoître („Let us put limits to the fury of invading everything, to know everything").[16] Knowledge here is not only power, but carries—also and especially in the will to knowledge of European science—the germ and the power of destruction and self-destruction within itself. De Pauw makes it clear that science has a hand in the conquest of the world in several ways, both as an actor and as a vehicle. And it is one of the fundamental instances that create and exercise power over meaning.

As a reaction to the developments in the second phase of accelerated globalization, de Pauw at the same time emphasized that peaceful coexistence under the

14 Ibid.

15 Pauw, Cornelius de: *Recherches philosophiques sur les Américains*, vol. I, p. a4v.

16 Ibid.

conditions of globality was always endangered from Europe, since every seemingly local dispute could ignite a world fire. For a long time now, the political and economic clashes of interests between Europeans had been so pronounced, especially in the colonial territories, that all it took was a spark to see the whole world go up in flames:

> [...] A spark of discord, for a few acres of land in Canada, sets Europe ablaze and ablaze; and when Europe is at war, the whole universe is at war: All the points of the globe are successively shaken as if by an electric power: The scene of massacres and carnage has been enlarged from Canton to Archangel; from Buenos-Aires to Quebec. The trade of Europeans having intimately linked the different parts of the world by the same chain, they are equally drawn into the revolutions and vicissitudes of attack and defence, without Asia being neutral, when some merchants have quarrels in America, for Beaver skins, or for Campèche wood.[17]

The same chain ("une même chaîne") connects the world and binds it together as a fetter for better or worse. Local disputes as well as conflicts in world trade, as repeatedly demonstrated in the eighteenth century, could suddenly escalate into military confrontations, which—as in the case of the Seven Years' War (1756–1763)—can justifiably be called World Wars.

After the publication of the second volume of the *Recherches philosophiques sur les Américains*, which de Pauw in many respects still sharpened, first in Berlin, but quickly also in various countries of Europe and above all overseas, a fierce controversy began, the first phase of which can with good reason be called the Berlin debate about the New World. For Antoine-Joseph Pernety, who as a former French Benedictine and ship's chaplain on Bougainville's 1763 voyage to the Malvinas had received the protection of Frederick II and an appointment as librarian to the 'Roi philosophe',[18] quickly came out vehemently against the New World in a lecture to the Berlin Academy on September 7, 1769. In the following year he published his *Dissertation sur* l'*Amérique et les Américains, contre les Recherches philosophiques de Mr. de P***.*[19] Thus had begun a dispute whose consequences were to become of worldwide significance far beyond the borders of Berlin, Prussia, and Europe. Prussia had become the center of a worldwide and long-lasting debate.

The weakness of Pernety's attack on de Pauw undoubtedly lay in the fact that he limited himself in many places to simply eloquently reversing de Pauw's theses, simply asserting the opposite and consequently countering a devaluation with an idealization of the indigenous population, which did not even stop at the recourse to the heterostereotype of the *bon sauvage*, of the 'noble savage'. Pernety's constant assertions that he had seen what he described with his own eyes were consequently only convincing to a limited extent. In a way, he provided the inverse representation of de Pauw's theses.

---

[17] Pauw, Cornelius de: *Recherches philosophiques sur les Américains*, vol. I, p. 90.

[18] See also Fontius, Martin: The place of the 'Roi philosophe' in the Enlightenment. In (ed.): *Frederick II and the European Enlightenment.* Berlin: Duncker & Humblot 1999, pp. 9–27.

[19] See Gerbi, Antonello: *La Disputa del Nuovo Mondo*, pp. 120–125.

No wonder, then, that de Pauw succeeded in capitalizing on the rising polemic by generously including Pernety's *dissertation* himself in a new three-volume edition in 1770 and combining it with a well-calculated counterattack. Certainly, one may recognize in Pernety "elements of a modern ethnological attitude",[20] among which one can certainly count "the critique of Eurocentrism; the importance of gathering data in the field and verifying sources; the recognition of the cultural diversity of the Other".[21] But his way of argumentation remained too much bound to the theses put forward by de Pauw, to which the librarian mostly only added other signs; and it did not help him much to present himself (with reference to his journey alongside Bougainville to the New World) as a knowledgeable traveller and eyewitness, while his opponent, as is well known, had never left Europe. An all-senses engagement with America counted for little in this debate: *a beau mentir qui vient de loin*. In the Berlin debate, in which Frederick II, as far as we know today, did not intervene directly and in which he must have enjoyed himself, Cornelius de Pauw sovereignly retained the upper hand and was more than ever able to assert with public effect that the cultivated peoples of Europe were infinitely superior to the "barbares de l'Amérique, qui ne savent ni lire, ni écrire, ni compter au-delà de leurs doigts"(barbarians of America, who can neither read nor write, nor count beyond their fingers).[22] It was the enthronement of the superiority of white Europeans over all other peoples of the world, accomplished by means of an Enlightenment philosophy.

Fundamental objections and contradictions against Cornelius de Pauw's theses, which went far beyond Pernety's horizon, were not long in coming in the eighteenth century.[23] In 1780, for example, the Jesuit Francisco Javier Clavijero, who had been expelled from New Spain, published an extensive history of ancient Mexico in his Italian exile,[24] in which he impressively demonstrated the diversity of the indigenous cultures of the Americas in the territory of present-day Mexico. In doing so, this Neo-Spanish author not only succeeded in countering the theses of Cornelius de Pauw, but also of Guillaume-Thomas Raynal or William Robertson, by including in his study a wide variety of American sources, including illuminated manuscripts and other indigenous documents; rather, he advanced a construction of the American past that, as—in the best sense—a finding and invention of a different origin, was also intended to make possible a different future for his American homeland. Pre-Columbian cultures no longer appeared here as negligible marginal phenomena of

[20] Mannucci, Erica Joy: The savage and the civilised: observations on a dispute between an enlightened writer and an illuminist. In: *Studies on Voltaire and the Eighteenth Century* (Oxford) 303 (1992), p. 384.

[21] Ibid.

[22] Pauw, Cornelius de: *Recherches philosophiques sur les Américains*, vol. 3, p. 7.

[23] See. on this in detail Ette, Ottmar: European Reflections on Two Phases of Accelerated Globalization in Cornelius de Pauw, Georg Forster, Guillaume-Thomas Raynal and Alexandre de Humboldt. In: Bancarel, Gilles (ed.): *Raynal et ses réseaux.* Texts collected and presented by Gilles Bancarel. Paris: Honoré Champion Editeur 2011, pp. 183–225.

[24] See Clavijero, Francisco Javier: *Storia Antica del Messico.* 4 vols. Cesena: Gregorio Biasani 1780.

human history, but confidently confronted a European antiquity as American antiquity. But the conquistadors who had invaded the cultures of 'ancient' America and sought to destroy their roots could now be considered barbarians.

It is revealing in this context that Clavijero not only reacted directly to the inferiority thesis of de Pauw's *Recherches philosophiques sur les Américains* as well as other works of the European Enlightenment with extensive research and publications, but at the same time also demanded to pay back the European Enlightenment philosophers "en la misma moneda" (in the same currency).[25] The mention of an African philosopher from Guinea, mentioned in the previous chapter, who was to describe Europe and its inhabitants from the opposite perspective, also belongs in this context. One could recognize in this a first step towards the reciprocity of arguments at eye level. For the Neo-Spanish Enlightenment philosopher suggested to write a work under the title *Investigaciones filosóficas sobre los habitantes del Antiguo Continente*, obviously based on de Pauw's work, that should follow "el mismo método de Pauw [sic!]" (the same method as de Pauw)[26] and (in painstaking philological detail) draw on all those writings that had brought negative aspects of the Old World into focus. Not only with regard to the phenomena of nature, but also those of culture, the most beautiful insights would then be offered, which, moreover, would be based on the writings of the ancient world itself:

> When he came to the article of vices, what an immense copy of materials would he not have for his work! What examples of vileness, perfidy, cruelty, superstition, and dissoluteness! What excesses in all sorts of vices! The history of the Romans alone, the most celebrated nation of the Ancient World, would furnish him with an incredible quantity of the most hideous evils.[27]

Clavijero had seen through de Pauw's method and at the same time taken it ad absurdum. With the New Spanish Jesuit, the Berlin debate entered a much wider context and opened up into a platform for discussion in which not only de Pauw and Pernety, Raynal and Robertson, but also the Neo-Spanish Dominican Servando Teresa de Mier, 'Le Philosophe la Douceur' alias Zaccaria de Pazzi de Bonneville,[28] Giovanni Rinaldo Carli, Delisle de Sales, Drouin de Bercy and many others took part. At the core of this was not least the question of on which side of the Atlantic lay the power over the meaning of the history of the New World and whether this meaning could be determined solely from a universe of texts—and thus without a sensual experience of America.

---

[25] Clavijero, Francisco Javier: *Historia antigua de México.* Prologue by Mariano Cuevas. Edition of the original written in Spanish by the author. Mexico: Editorial Porrúa ^{7}1982, p. 524.

[26] Ibid.

[27] Ibid.

[28] See Zantop, Susanne: *Colonial Fantasies*, p. 47.

In the Universe of the Senses

Was it a coincidence that with the Berlin debate in 1769 the actual 'hot phase' of the international dispute about the New World began in the Prussian capital of all places, which seemed rather provincial compared to Madrid and Lisbon, but above all Paris, Amsterdam and London? Certainly, with Frederick II, a personality had become king in Prussia who, only a few weeks after his accession to the throne with the invasion of Silesia in December 1740, left no doubt that he wanted to write a new score for the concert of the great powers. But in the second phase of accelerated globalization, Prussia was not a *global player* and had no overseas possessions; more precisely, the territorially torn kingdom was—as we saw in the previous chapter—*no longer* a colonial power.

Contrary to popular belief, Brandenburg and Prussia had previously engaged in colonial activities and attempted to secure their share in the expansion of Europe. The Great Elector had already attempted to make a profit in the lucrative colonial business between the Old and New Worlds, albeit with varying degrees of luck and skill. It was not without reason that Frederick William I (1640–1688) had studied in Amsterdam and had been able to observe at close quarters how maritime trade and colonial policy could catapult even a small country into the role of a leading international trading power.[29] In 1647, a former Dutch admiral, Arnoult Gijsels, acting as advisor to the Elector, proposed the establishment of a trading company under Brandenburg's leadership. Parallel to this, negotiations were conducted with Denmark on the sale of the Danish base at Tranquebar.[30]

After various setbacks, it was once again a Dutchman, the shipowner Benjamin Raule, who helped Brandenburg to build up a fleet and was appointed 'Chief Director in Maritime Matters' in 1677 and later Minister for Maritime Affairs and the Colonies.[31] Finally, in 1680, two expeditions set sail under the Brandenburg flag. In the immediate vicinity of Dutch and British settlements, Brandenburg's first colonial base in Africa was established, the Großfriedrichsburg settlement, which controlled a coastal strip of what is now Ghana, some 50 km in length. The second Brandenburg fleet, meanwhile, succeeded in bringing up and capturing the Spanish ship 'Carolus Magnus' off Ostend, which under the name 'Margrave of Brandenburg' enjoyed a second career as the flagship of the Brandenburg fleet.[32] The Brandenburgers' first privateering voyages in the Caribbean were followed in 1682 by the founding of the 'Brandenburg African Company', which was heavily manned by Dutchmen and partly by Huguenots, and was renamed the 'Brandenburg African American Company' in 1692. Emden in East Frisia was declared by contract to be

[29] See Lennert, Gernot: Kolonisationsversuche Brandenburgs, Preußens und des Deutschen Reiches in der Karibik (Colonization attempts by Brandenburg, Prussia and the German Empire in the Caribbean). In: Carreras, Sandra/Maihold, Günther (eds.): *Preußen und Lateinamerika. Im Spannungsfeld von Kommerz, Macht und Kultur.* Münster LIT-Verlag 2004, p. 11.

[30] Ibid.

[31] Ibid., p. 12.

[32] Ibid., p. 12 f.

the main port of the Brandenburg fleet: The small Brandenburg had advanced to become a colonial power that hoped to make profits from colonial goods, but above all from the lucrative transatlantic slave trade.

The hopes for profits as well as for an increase in power seemed to be fulfilled at first. Thus, as early as 1686, a first Brandenburg ship with 450 slaves from Großfriedrichsburg headed for the island of St. Thomas, which was shared with Denmark. This Brandenburg settlement in the Caribbean soon included three hundred Europeans and several hundred slaves. About thirty thousand slaves were taken from Africa to the West Indies, shipped and sold under Brandenburg flags and inhumane conditions.[33] Although Elector Frederick III, who succeeded the 'Great Elector' in 1688 and from 1701 bore the precarious title of King in Prussia as Frederick I, attempted to continue his predecessor's colonial ambitions and activities, the 'Brandenburg-African-American Company' nevertheless sank into debt and was finally bought out by Frederick William I., who had ruled since 1713, at a time when colonial business appeared less and less attractive in the concert of European powers before the second phase of accelerated globalization, which had not yet begun to emerge. In 1738 the last property of the Company was auctioned off.[34] The colonial history of Brandenburg-Prussia was thus for the time being history.

Barely two years later, Frederick II ascended the Prussian throne after the death of his hated father. A 46 year reign began, in the course of which the king remodelled Prussia to his own liking and undoubtedly—to use the words of Richard von Weizsäcker on the occasion of the bicentenary of Frederick's death in 1986—left behind as his legacy an "unsentimental polity with a strong character and a willingness to reform",[35] but certainly also a country with a great power policy that was unpredictable for a long time. But it was not until the newly founded German Empire, in the third phase of accelerated globalization, that it was again to pursue a colonial policy in the true, expansive sense, and with the infamous Berlin Africa Conference in 1884 it was to send a clear signal of its claims to power. Prussia and Germany had again become highly aggressive colonial powers that tried to make up for their missing colonial history in an exploitative way.

It cannot be ruled out with certainty that Frederick II's Prussian royal court was sometimes caught up in colonial fantasies,[36] even if, unlike London and Paris, the European metropolises of the second phase of accelerated globalization, the Prussian capital could no longer pursue any direct colonial interests of its own. The unpredictable Prussian monarch, however, not only discussed in his round table

[33] See ibid., p. 16 f.

[34] Ibid., p. 23.

[35] Weizsäcker, Richard von: Friedrich der Große: Geschichte und Mythos (Frederick the Great: History and Myth). In (this): *Demokratische Leidenschaft. Reden des Bundespräsidenten.* Edited and introduced by Eberhard Jäckel. Stuttgart: Deutsche Verlags-Anstalt 1994, p. 76.

[36] See on this from the perspective of 'postcolonial studies' the work of Zantop, Susanne: *Colonial Fantasies. Conquest, Family, and Nation in Precolonial Germany, 1770–1870.* Durham/London: Duke University Press 1997, esp. pp. 46–65.

with Voltaire and others what was just beginning to emerge as world politics, but also displayed a sustained interest in that thinker of globality whom we recognized in Cornelius de Pauw in the first part of these reflections. It is not without reason that his writings were published in Berlin in the sixties and seventies.

Thus de Pauw also became "reader to Frederick the Great, in whose circles the presence of the free-spirited Abbé was an indispensable piece of decoration".[37] Not only during de Pauw's first stay in Berlin and Potsdam between 1767 and 1768, but also during the second one between 1775 and 1776, the monarch made an effort to find the Abbé and introduced him as a "new acquisition" to his illustrious round table (which, admittedly, had seen better days).[38] Even if we know little in detail about the relations between Frederick II and his reader and the name of de Pauws is mostly conspicuous by its absence even in the standard works on Frederick the Great, because today in the German-speaking world one seems to have no idea of the enormous international reputation of the author of the *Recherches philosophiques sur les Américains*, one can certainly assume that the Prussian king could not have been uninvolved in that Berlin debate that flared up on September 7, 1769. September 1769 the Berlin debate which broke out between his librarian Pernety and his (former) reader de Pauw at that Academy of Sciences Frederick had re-founded in 1744 as patron of the sciences.

Conversely, it cannot have escaped Cornelius de Pauw's notice that his royal host had dealt with an American subject a decade and a half earlier in a libretto that has only recently been taken more seriously by scholars, and which, from today's perspective, we can thematically assign to the first phase of accelerated globalization. Thus it seems like an ironic wink when the Abbé, born in Amsterdam in 1739 and not yet 30 years old, brands the conquest of the Aztec empire by the Spaniards in his *Recherches philosophiques* as the raid of four hundred murderers and bandits,[39] whom Hernán Cortés had gathered around him as arbitrarily as he had greedily. Certainly, he was not unaware of the passage by the royal librettist in which the Aztec ruler Montezuma—in the Italian translation by Frederick's court poet Giuseppe Tagliazucchi[40]—disregards the warnings of his general Pilpatoè about the danger of the approaching Spaniards:

---

[37] See. Beyerhaus, Gisbert: *Abbé de Pauw und Friedrich der Große, eine Abrechnung mit Voltaire (Abbé de Pauw and Frederick the Great, a Reckoning with Voltaire)*, p. 467.

[38] See Beyerhaus, Gisbert: Abbé de Pauw und Friedrich der Große, eine Abrechnung mit Voltaire (Abbé de Pauw and Frederick the Great, a Reckoning with Voltaire). In: *Historische Zeitschrift* (Munich/Berlin) 134 (1926), p. 468.

[39] Pauw, Cornelius de: *Recherches philosophiques sur les Américains*, vol. 1, p. 58.

[40] See on this Polzonetti, Pierpaolo: Political and Operatic Reforms: of Frederick the Great. In (this): *Italian Opera in the Age of the American Revolution.* Cambridge: Cambridge University Press 2011, p. 109. On the history of the reception of Montezuma and Moctezuma in Spanish theater, see, among others, Pérez-Amador Adam, Alberto: La constitución del modelo dramático del personaje de Moctezuma II en el teatro español de finales del siglo XVIII e inicios del siglo XIX (The constitution of the dramatic model of the character of Moctezuma II in the Spanish theater of the late 18th and early 19th centuries.). In: Floeck, Wilfried/Fritz, Sabine (eds.): *La*

> Heavens! To what extent does your fear
> Your heart can't fail you To the defense
> Of Mexico be necessary
> An entire Army
> Against three hundred errant fugitives
> Puked up by the sea on these shores?[41]

But these three hundred marauding fugitives, who sometimes appear as vagabonds,[42] will defy all warnings to advance into the heart of Montezuma's empire and take the place of the rightful ruler of the Aztecs with all brutality and greed. By this highly risky seizure of power they certainly did not become men of honour—not only in Frederick's eyes.

This is the plot of the opera *Montezuma*,[43] whose libretto Frederick II had been working on since the late summer of 1753, and which has recently come to the public's attention again.[44] Is it a coincidence that in the carnival of the year 1755 it had its premiere at the Royal Opera House on January 6? Is it a coincidence that it premiered in the Royal Opera House on Unter den Linden in the carnival of 1755 on January 6, the same Epiphany on which the Catholic Monarchs Fernando and Isabel completed the conquest of the Nasrid Empire with their ceremonial entry into Granada in 1492, ending centuries of Arab rule on the Iberian Peninsula and turning the *Reconquista* into a future *Conquista* of countries outside Europe—the Moorish empires south of the Strait of Gibraltar were initially in view?

What is certain is that the Prussian monarch was aware from the outset of the political implications of what is today certainly his best-known opera project; indeed, he was concerned with a political message that could be conveyed as efficiently and sustainably as possible, as he informed Francesco Algarotti early on in a letter probably dated October 1753.[45] In it, he reported to his close friend on his work on the libretto:

---

representación de la Conquista en el teatro español desde la Ilustración hasta finales del franquismo. Hildesheim/Zürich/New York: Georg Olms 2009, pp. 161–177.

[41] [Frederick II./Tagliazucchi, Giuseppe] *Montezuma. Tragedia per música.* To be performed in the Royal Theater of Berlin by Order of His Majesty the King in the Carnival of the MDCCLV, Berlin: Printed by Haude and Spener with license of His Majesty 1755, p. 32; there is also the German-language version: "Wie sehr, oh Himmel! kann dir deine Furcht / Den Muth benehmen… Is then / Against three hundred erring fugitives, / Which the sea has spewed on this shore, / To defend Mexico, a whole army necessary?" (p. 33).

[42] See Klüppelholz, Heinz: Die Eroberung Mexikos aus preußischer Sicht—Zum Libretto der Oper 'Montezuma' von Friedrich dem Großen (The Conquest of Mexico from a Prussian Perspective—On the Libretto of the Opera 'Montezuma' by Frederick the Great). In: Gier, Albert (ed.): *Oper als Text: Romanistische Beiträge zur Libretto-Forschung.* Heidelberg: Winter 1986, p. 82.

[43] Performances of *Montezuma* have recently taken place at the Mannheim Opera (26 March 2010), at the Hamburg Kulturfabrik Kampnagel (10 June 2010), in Mühlheim an der Ruhr (1 July 2010), at the Schlosstheater in Potsdam (21 January 2012) and a concert performance at the Staatsoper Unter den Linden (alternative venue Schiller Theater, 28 January 2012).

[44] See Henze-Döhring, Sabine: *Friedrich der Große: Musician and Monarch.* Munich: C.H. Beck 2012, p. 85.

[45] See on this dating ibid., p. 217.

> It's Montézuma. I have chosen this subject, and now I am adapting it. You can see that I will be interested in Montézuma, that Cortès will be the tyrant, and that consequently we will be able to unleash, in music itself, some lardon against the barbarity of the R.C. But I forget that you are in a country of the Inquisition; I apologise for this, and I hope to see you soon in a heretical country, where opera itself can serve to reform morals, and destroy superstition.[46]

That Friedrich in his planned opera consequently aims at the *religion catholique*, at Catholicism and its machinery of oppression, is from the beginning of the project thus just as clear as the fact that his opera *Montezuma* is no less about the question of power than about the (sub)divorce between barbarism and civilization. The fact that he intends to do this *en musique même*—which should not be translated, as has often been done, as 'even in the music', but rather 'in the music itself'—sheds a telling light on his understanding of a collaboration of the arts, which aims at a collaboration of the senses. And this orchestration of the senses, as will be explained below, is linked in the medium of opera to the question of power in an intimate, highly sensual way.

On April 16, 1754, Frederick II wrote from Potsdam to his favorite sister and proven musical advisor, Margravine Wilhelmine of Bayreuth, transmitting to her the libretto written in French with the following words:

> I take the liberty of laying at your feet a Mexican who is not yet quite cultured. I have taught him French; now he must learn Italian. But before I subject him to this trouble, I beg you very much to tell me frankly your opinion, and whether you think it worth while. The majority of the arias are not to receive a da capo; only two arias of the Emperor and two of Eupaforice are intended for this. I do not know how you will find the whole, the sequence of scenes, the dialogue, and whether the plot is interesting. But as the matter is not hurried, I could easily change what you find fault with. The effect of the play could even be easily tested. As you have a wonderful French theatre company, they need only perform it in your room, even if the parts are only read off.[47]

---

[46] Friedrich II: *Correspondance de Frédéric Second, Roi de Prusse, avec le Comte Algarotti.* To serve as a sequel to the editions of the posthumous works of this Prince. Berlin: George Gropius, Bookseller 1887, p. 98.

[47] Letter from Frederick II to Margravine Wilhelmine of Bayreuth, April 16, 1754; I would like to thank Ruth Müller-Lichtenberg for kindly transmitting the correspondence between the two siblings. In French Frederick the Great the passage reads: "Ma tres chere Sœur. Je prens la liberté de metre a Vos pieds un Mexsiquain qui n'est pas encore toutafait de crassé je lui ai apris a parler français, il faut apresent quil aprene L'Italien, mais avans que de Lui donnér cette pene je vous suplie de me dire Naturelement Votre sentiment et si Vous Croyéz quil merite qu'on Lui donne ce soin; La plusparc des airs sont faits pour ne point etres repetéz il n'y a que deux airs de L'Empereur et deux d'eupaforis qui sont destinéz pour L'Etre, je ne sai Coment Vous trouveréz le tout ensemble L'enchainement des Scenes, Le Dialogue et L'Interet que j'aurois Voulû y faire regnér, mais Comme rien ne presseje pourai changér facilement ce que Vous y trouverais a redire; il seroit meme facile de jugér de L'Efet que le Spectacle peut produire, Vous avéz une admirable troupe françaisse il n'yauroit qu'a la lui faire representér dans Votre chambre qund meme chaqun ne ferait que Lire son role". (My dearest sister. I take the liberty of placing at your feet a Mexican who is not yet fully grown up. I have taught him to speak French, but now he must learn Italian, but before giving him this task I beg you to tell me your feelings and whether you think he deserves to be taught; Most of the arias are made not to be repeated, there are only two of the Emperor's arias and

From the very beginning, Frederick II relied on the effect of the word as well as on the effect of the music, for the execution of which he had already developed clear guidelines—especially the turning away from dacapo arias and towards the two-part lyrical form of the cavatina.[48] The sister's reaction was not long in coming: In a letter to her brother on April 25, 1754, she expressed herself "delighted,"[49] a delight that later led to her plan to perform the work at the Margravial Opera House in Bayreuth.

In the subsequent correspondence, Frederick's not infrequently enthusiastic comments finally culminate in the letter of January 11, 1755, in which the king self-deprecatingly but not without enthusiasm reported to his sister on the first performance of 'Montezuma' in the Royal Opera House on Unter den Linden:

> The decorator and the costume designer got the poor lyricist out of the affair. Mainly two pistol shots have won extraordinary applause. The Astrua played the last scene with admirable pathos and Graun outdid himself in the music.[50]

Ruth Müller-Lindenberg has convincingly placed the exhibition she conceived and opened in January 2012 at the Berlin Museum of Musical Instruments under the title 'Frederick's *Montezuma*' entirely under the sign of all those senses that Frederick II was quite consciously intended to address. For it is not without reason that the king's letters speak not only of his French libretto and its Italian translation or of the

---

two of the Eupaforis which are destined to be, I do not know how you will find the whole thing together, the sequence of the Scenes, the Dialogue and the Interest which I would have liked to make prevail, but as there is no hurry I can easily change what you find to be wrong with it would be easy to judge the effect that the Spectacle can produce, You have an admirable French troupe, all you have to do is to have it performed in Your room, and even then each one would only be reading his part) (GStA BPH Rep. 47 N° 305–310, fol. 19 r).

[48] See on this from an artistically quite critical view of the opera Detering, Susanne: Friedrich II: Camouflage und Selbstverteidigung im Bild Montezumas. In (this): *Kolumbus, Cortés, Montezuma: Die Entdeckung und Eroberung Lateinamerikas als literarisches Sujet in der Aufklärung und im 20. Jahrhundert.* Weimar: Publishing House and Databases for the Humanities 1996, p. 196. From a musicological perspective, See Henze-Döhring, Sabine: *Friedrich der Große: Musician and Monarch*, p. 85 ff.

[49] Letter from Wilhelmine of Bayreuth to Frederick II, April 25, 1754. In the French original: "J'ai eu L'honneur de recevoir votre Heros Mexiquain j'en ai étée enchantée. Les Sentiments, Les positions et la Conduite de la piece, touts cella dis je est exelant je faits ecrire les Rôles pour Les faire distribuer aux Comediens, et je conte que dans 3 semaines La piece sera jouée." ("I had the honor of receiving your Mexican hero and was delighted. The feelings, the positions and the conduct of the play, all of which I say are excellent, I have the roles written down for distribution to the comedians, and I count that in 3 weeks the play will be performed"). (GStA BPH Rep 46 W 17, vol. III.5 (1754), fol. 11 r/v).

[50] Letter from Frederick II to Wilhelmine of Bayreuth on January 11, 1755. In the French original: "Le decorateur et le tailleur ont tiré Le pauvre Auteur d'aiffaire [!] surtout deux Mauvais Coups de pistolet ont etéz extremelyment aplaudi. L'astrua a joué la Derniere Scene avec un patetique admirable et Grauen s'est surpassé en Musique." ("The decorator and the tailor have pulled the poor author [!] especially two bad pistol shots have been extremely applauded. L'astrua played the last Scene with an admirable pathetic and Grauen surpassed himself in music" (GStA BPH Rep. 47 N° 305–310, fol. 71 v).

music of his court conductor Carl Heinrich Graun, but also of the stars of the performance, which included, in addition to the highly paid castrati, the no less remunerated prima donna Giovanna Astrua, the costume designers and theatre architects, above all Giuseppe Galli Bibiena, and all those artists and craftsmen who are responsible for the *special effects*—from the pistol shots to the pyrotechnics used on stage.

The young crown prince probably first experienced the power of the interaction of all these forces at first hand during his visit to the Royal Electoral Saxon Court in Dresden in January and February 1728, when he experienced the "dizzying contact with a courtly scenario, which in its radiant splendour, its enticing, playful lightness and the breathless succession of comedies and ballets, of gala banquets and redoubts, masquerades and fireworks as a grandiose dazzling work beguiling the senses".[51]

In her groundbreaking exhibition, Ruth Müller-Lindenberg thankfully did not shy away from having an olfactory designer recompose that scent that must have formed in a closed opera house from the candles and chandeliers, the perfumes used and the sweat, but also the fireworks and gunpowder vapor during the performance. It is no coincidence that she gave her introduction to the performance catalogue the title 'Power and Senses?' and prefaced her exposition with a quotation from Julius Bernhard von Rohr's 1733 *Introduction to the Ceremoniel-Wissenschaft der großen Herren*:

> If the subjects are to recognize the king's majesty, they must understand that he has the highest authority and power, and accordingly they must arrange their actions in such a way that they have reason to recognize his power and authority from them. The common man, who is attached only to the external senses and makes little use of reason, cannot quite imagine what the king's majesty is, but by the things that come into his eyes and move his other senses, he gets a clear idea of his majesty, power, and authority.[52]

In this sense, the spectacle of opera was a spectacle of power and violence, aimed above all at one thing: Achieving the greatest possible effect on the audience. In Friedrich's *Montezuma*, therefore, it is not only on the level of content, but no less on the level of expression, that which overwhelms the senses, eludes mere reason and cannot be unfolded by the poor librettist in a universe of texts alone. The pistol shots on the open stage, the fireworks in a daring stage architecture, the exotic costumes and elaborate sets are not committed to an *effet de reel*, but to an *effet de pouvoir* which, by virtue of its power over the senses, is able to bring the sense of power under the skin as artfully as it is 'naturally'. When you say opera, you are certainly saying power—in the spirit of Frederick II, who had the foundation stone of his great opera house laid the year after he came to power. The opera house on Unter den Linden, the first of Frederick's many major building projects that were to decisively shape the residential cities of Berlin and Potsdam, impressively

[51] See Kunisch, Johannes: *Friedrich der Große. Der König und seine Zeit.* With 29 illustrations and 16 maps. Munich: German paperback Publisher ^{2}2010, p. 22.

[52] Cited in Müller-Lindenberg, Ruth: Power and senses? On the Concept of the Exhibition (introduction to the as yet unpublished exhibition catalogue 2012).

demonstrates how the Prussian king knew how to combine the promotion of the arts with the promotion (and reflection) of his own power.

The intended confusion of the senses consequently aims at a seduction of the senses, the aesthetic power of which is not without repercussions for the relationship to the political power of the monarch. This does not at all mean that reason is thereby largely eliminated or completely overwhelmed in the sense of von Rohr's ideas. For reason too—only not 'mere', naked reason—is drawn into the play of confusions and set in motion with the help of words, which we can view in isolation in the libretto as in a reading drama. This happens in Friedrich's opera *Montezuma* right from the first scene of the first act, which from the beginning links everything to the question of power and its appropriate use.

This question, of course, was no less central to the political-philosophical writings attributed to Frederick II. After reading Machiavelli's *Il Principe*, in a letter to Voltaire dated March 31, 1738, he already communicated his intention to carefully refute this work written in the spirit of medieval princely mirrors, which Frederick read in French translation.[53] Voltaire encouraged the Prussian crown prince in his intention and spoke of how this work could one day become the "catechism of kings and their ministers".[54]

Frederick probably began writing it in May 1739 and probably completed his overall conception in November of the same year, subsequently revising his work several more times. On April 26, he sent his writing to Voltaire in the hope that his text would certainly win under the pen of the French philosopher. Thus two versions were produced, which were published independently of each other at The Hague in 1740: First, Frederick's *Examen du Prince de Machiavel, avec des notes historiques et politiques* and second, the *Anti-Machiavel, ou Essai de critique sur le Prince de Machiavel, publié par Mr de Voltaire*.[55]

Even if the complex history of the origins of this work cannot be pursued further here, it is nevertheless decisive for our question, which focuses on Prussia, that Frederick emphasized early on and right at the beginning of this 'Réfutation du Prince de Machiavel', which was initially published anonymously, his view that Machiavelli's *Il Principe* was "one of the most dangerous of all books of world circulation".[56] And with the same thrust he pugnaciously elaborated: "Machiavell

---

[53] See Kunisch, Johannes: *Friedrich der Große. Der König und seine Zeit*, p. 124.

[54] Quoted from ibid., p. 125.

[55] Quickly appeared further and expanded editions, such as that published the following year in Amsterdam by Frederick II: *Anti-Machiavel, ou Essai de critique sur le Prince de Machiavel, publié par Mr de Voltaire*. New Edition, with added variations of the London Edition. Amsterdam: at Jaques La Caze 1741. The complex history of the creation and dissemination of this writing cannot be presented here; see the critical edition of the *Anti-Machiavel*. Ed. par Werner Bahner and Helga Bergmann. *Les Oeuvres Complètes de Voltaire. Vol.* 19. Oxford: the Voltaire Foundation 1996. There you will also find the text of the French edition that Frederick had consulted.

[56] Quoted here from the widespread version in Frederick the Great: Antimachiavel. In (this): *Historische, militärische und philosophische Schriften, Gedichte und Briefe*. With illustrations by Adolph von Menzel. Ed. by Gustav Berthold Volz. Cologne: Anaconda 2006, p. 331. The French original of this writing is quoted from Frederick the Great: *Anti-Machiavel ou Examen du Prince*

planted the germ of corruption in state life and undertook to destroy the precepts of sound morality."[57] Unlike the writings of the philosopher Spinoza, Machiavelli's considerations of power theory in *The Prince* had not led (merely) to the "aberrations of thought," but had unfolded a (doubtless fatal) "significance for life itself."[58] Hence the young crown prince and would-be king's plan to take up the "defense of humanity against a monster who wants to corrupt it."[59] A rejection of the (exercise of) power, even its swan song?

Not at all. If one reads Frederick's *Antimachiavel* from the perspective of the opera written a decade and a half later and its libretto, there can be no doubt that Frederick's Hernán Cortés represents the embodiment of all those qualities that characterize Machiavelli's prince: "greed", "caprice", "ambition", "treachery", "cruelty"—in short, the image of a "political monster[s]".[60] Even in his poem 'Death of the Father', dated February 26, 1740 and dedicated to Voltaire, Frederick will once again recall "Machiavelli's spirit of sacrilege" immediately before his accession to the throne and engage in a skillful game of dissimulation, as if he hesitated to reach for power now.[61] It was precisely in the not infrequently violent disputes with his father that the heir to the throne had acquired those "arts of disguise

de Machiavel. Corrected for the most part according to the original manuscript of Frederick II. With an introduction and historical notes. Hamburg: Frédéric Perthes 1834, p. 4: "J'ai toujours regardé le Prince de Machiavel comme un des ouvrages les plus dangereux qui se soient répandus dans le monde." ("I have always looked upon Machiavelli's Prince as one of the most dangerous works that has ever spread through the world").

[57] Frederick the Great: Antimachiavel, p. 331. In the original French, the passage reads, Frederick the Great: *Anti-Machiavel ou Examen du Prince de Machiavel*, p. 4: "Machiavel corrompit la politique, et entreprit de détruire les préceptes de la saine morale." ("Machiavelli corrupted politics, and set out to destroy the precepts of sound morality").

[58] Frederick the Great: Antimachiavel, p. 331. In the French original: Frederick the Great: *Anti-Machiavel ou Examen du Prince de Machiavel*, p. 3: "erreurs de spéculation", "regardoient la pratique". ("errors of speculation", "look at the practice").

[59] Friedrich the Great: Antimachiavel, p. 331. In the French original: Friedrich the Great: *Anti-Machiavel or Examination of Machiavelli's Prince*, p. 4: "J'ose prendre la défense de l'humanité contre un Monstre qui vent la détruire." ("I dare to take the defense of mankind against a Monster who wants to destroy it").

[60] Friedrich the Great: Antimachiavel, p. 332. In the French original: Friedrich the Great: *Anti-Machiavel or Examen du Prince de Machiavelli*, p. 6: "[…] et quelle situation déplorable que celle des peuples, lorsqu'ils doivent tout craindre de l'abus du pouvoir souverain, lorsque leurs biens sont en proie à l'avarice de leur prince, leur liberté à ses caprices, leur repos à son ambition, leur sûreté à sa perfidie, et leur vie a sa cruautés. C'est lá le tableau d'un empire où régneroit un monstre politique tel quel Machiavel prétend le former " ("[…] and what a deplorable situation it is for the people, when they have to fear everything from the abuse of sovereign power, when their goods fall prey to the avarice of their prince, their liberty to his caprices, their rest to his ambitions, their safety to his perfidy, and their lives to his cruelties. This is the picture of an empire where a political monster such as Machiavelli pretends to form would reign").

[61] Friedrich the Great: Father's death. In (this): *Historische, militärische und philosophische Schriften, Gedichte und Briefe*, p. 545. In the French original: Friedrich the Great: *Correspondance de Frédéric avec Voltaire*. Volume I. From the first letter of Frederick to his accession. (August 8, 1736–June 4 or 5, 1740). In: Preuss, Johann D (ed.): *Euvres de Frédéric le Grand*. Volume XXI. Berlin: Royal Printing House (R. Decker) 1853, p. 398: "Dans ce terrain scabreux, raboteux,

bordering on deviousness", "with which he first knew how to hoodwink his father and then his opponents in the concert of powers".[62]

But Frederick did not hesitate: He seized the opportunity. However, against the image of despotism and tyranny, he had impressively placed another image of a ruler in his *Antimachiavel*: "Thus it amounts to this, that the ruler, far from being the unrestricted master over his subjects, is only their first servant, the instrument of their happiness, as those are the instruments of his glory."[63] The philosopher on the throne: Which role, which score did Frederick play with greater devotion, with greater fervor?

It is with this image of the servant ruler that Frederick II opens his opera *Montezuma* in 1755, in the first scene of which the Aztec ruler can look with pride on a reign that has been happy and glorious for his subjects as well as for himself. Against the backdrop of three large palm walks in his imperial garden, he addresses Tezeuco, a servant from his numerous entourage:

> Yes, my Tezeuco, Mexico is happy.
> Fruit of that freedom, which, unique
> To prudence, to frenzy alone
> Of the laws, which I myself
> I am the first to observe, my people.
> Of stable joy, and beautiful rest abounds,
> And my power upon his love is founded.[64]

Thus the commonwealth of the Mexica presents itself in the first recitative of their ruler in obvious agreement with that ideal image which the future Prussian king had sketched out so hopefully for all in the *République des Lettres* in his first political-philosophical treatise. Did Frederick not speak there—as if he were speaking of a *contrat social*—of those peoples "who have set a lord over themselves merely so that he may be their shield and protection", when they had "only submitted to him on this condition"?[65]

---

difficile / De machiavélisme infecté!" ("In this scabrous, rough, difficult terrain / Of Machiavellianism infected!")

[62] Kunisch, Johannes: *Friedrich der Große. Der König und seine Zeit*, p. 25.

[63] Friedrich the Great: Antimachiavel, p. 332. In the French original: Friedrich the Great: *Anti-Machiavel or Examen du Prince de Machiavelli*, p. 10: "il se trouve que le souverain, bien loin d'être le maître absolu des peuples qui sont sous sa domination, n'en est en lui-même que le premier domestique." ("it so happens that the ruler, far from being the absolute master of the peoples under his rule, is in himself only the first servant").

[64] *Montezuma. Tragedia per musica*, p. 10. "Ja, mein Tezeuco, Mexico ist glücklich. / Dieses ist die Frucht jener Freiheit, welche auf / Vernunfft gegründet, nur der Herrschafft solcher / Gesetze unterworfen ist, die ich selbst zu erst beobachte. / Mein Volk genießt in Ueberfluß eines / Festgegründeten Glücks und einer holden Ruhe, / Und meine Macht gründet sich auf dessen Liebe". ("Yes, my Tezeuco, Mexico is happy. / This is the fruit of that freedom which is founded on / reason, subject only to the rule of such / laws as I myself first observe. / My people enjoy in abundance a / Firmly founded happiness and a fetching peace, / And my power is founded on its love") (p. 11).

[65] Friedrich the Great: Antimachiavel, p. 332. In the French original: Friedrich the Great: *Anti-Machiavel or Examen du Prince de Machiavelli*, p. 11: "qui se sont donné des souverains pour

Frederick plays with the images of power. While King Ferdinand of Aragon was for Machiavelli the epitome of a prince acting with great skill, the same Ferdinand who conquered Granada and, together with Queen Isabel, directed Spain's impulse for power towards the coasts of the New World, is subjected to harsh Enlightenment criticism for having used religion "as a cloak" for his plans and abused the "pious faith of the people" for his "hypocritical actions".[66]

But with it that European monarch is punished who not only knew how to rule in alliance with the Church and the 'Holy Inquisition'—which Frederick derided in his already mentioned letter to Algarotti—but also decisively shaped the first phase of accelerated globalization politically. Thus in the *Antimachiavel* those enemies of reason are already early identified for whom the Spaniards in the French-influenced eighteenth century were taken without much ado. Spain and the Catholic Church, on both sides associated with the *Santa Inquisición*: Thus, in the political-philosophical treatise, those universally hated disruptive factors for a more just course of the human race are identified early on, against which Frederick's opera *Montezuma* attempts to burn off its sensuous fireworks.

Before his atrocities and massacres of an innocent population in the seventh scene of the second act, did not Cortés, like a second Ferdinand, still try to cover up his shameful actions with a reference to the Christian religion, and call the Mexica he slaughtered barbarians with reference to their human sacrifices?We by law abhor the wicked idolater,

> Who offered human victims to barbaric gods.
> More than conquering,
> Let us make our God known to you,
> And establish between you that perfect
> Religion, which to this Nume is accepted.[67]

Who the 'true' barbarians and who the 'civilized'[68] are in this scene from the year 1521 is laid bare for the audience at the Linden Opera. The extent to which the young crown prince was already preoccupied with the questions and challenges of

---

qu'ils les protègent", "qui ne se sont soumis qu'à cette condition." ("who have given themselves rulers to protect them," "who have submitted only on this condition").

[66] Friedrich the Great: Antimachiavel, p. 332. In the French original: Friedrich the Great: *Anti-Machiavel or Examen du Prince de Machiavelli*, p. 169: "un voile pour couvrir ses desseins," "son hypocrisie la crédulité des peuples au profit de son ambition." ("a veil to cover his designs," "his hypocrisy the gullibility of the people for the benefit of his ambition").

[67] *Montezuma. Tragedia per musica*, p. 70. "Unser Gesetz will, daß wir die Götzen-Diener verabscheuen, / Welche barbarischen Göttern menschliche Opfer bringen. / Wir sind weniger darauf bedacht, Länder zu erobern, / Als vielmehr hier unsern Gott bekannt zu machen, / Und unter euch denjenigen vollkommenen / Gottes-Dienst zu stiften, der diesem Gotte angenehm ist. "Our law wants us to abhor the idol-servants, / Who bring human sacrifices to barbarian gods. / We are less anxious to conquer lands, / Than to make our God known here, / And to establish among you that perfect / Service of God which is pleasing to this God" (p. 71).

[68] On this line of conflict of European reflection on the encounter with non-European cultures, see Bitterli, Urs: *Die 'Wilden' und die 'Zivilisierten'. Grundzüge einer Geistes- und Kulturgeschichte der europäisch-überseeischen Begegnung.* Munich: C.H. Beck'sche publishing house 1976.

the globalization process not only of the sixteenth century, but also of his own time, can be seen—along with many other writings from his pen—in various passages of his *Antimachiavel*, where, for example, it is a question of the strategies of France to cross the trade routes of England and Holland, the "two most prosperous and richest countries in the world",[69] and to use them to its own advantage.

Frederick was well aware of the fact that it had long since ceased to be Spain and Portugal, but Holland, and above all England and France, who were acting with increasing dynamism on a global scale and were consequently able to unleash a real great power policy that he himself would have been unable to counter with anything equivalent. The global dimension of this action undoubtedly seemed to him most promising in the field of trade, without, however, seeing an epoch of world peace emerging from what was still beginning to be called world trade in the eighteenth century.

The prospective monarch was certainly well aware of the colonial activities of Brandenburg and Prussia, since the last remnants of the former company of Prussian overseas trade—as we have seen—had been dissolved and liquidated only a year before the publication of his *Antimachiavel*. Did he see in a possible revival of Prussian colonial activities a field of political action of his own? Nothing in his treatise, nothing in his opera, nothing in the research devoted to Frederick points directly to this.

And yet, against this background, it is significant to understand that Frederick II's artistic preoccupation with the first phase of accelerated globalization in his opera *Montezuma* is shaped by its own horizon of experience and expectation, which would not have been possible without the onset of the second phase of accelerated globalization. With the power of art, Frederick spans a transmedial experimental space that does not serve any mimesis of concrete historical power, but rather the aesthetic testing of abstract notions of power on a global scale.

The Sense of Power and the Power of the Senses

One may understand the recitative immediately before the first aria of Montezuma as a direct reminiscence of *Antimachiavel*, when the Aztec ruler responds to the hymns of praise of his servant Tezeuco with turns of phrase which, in literally condensed form, project his ideal image of a ruler into the enclosed space of a Prussian opera performance with hyper-clear brightness:

> Do not raise so much
> Yes, weak virtues. I never do anything else
> That my duty? I'm asking you, man,
> Not praise, but advice. And merit in us

[69] Frederick the Great: Antimachiavel, p. 339. In the original French: Frederick the Great: *Anti-Machiavel ou Examen du Prince de Machiavel*, p. 174: "l'Angleterre et la Hollande, ces deux pays les plus beux et les plus riches du monde." ("England and Holland, those two most beautiful and wealthy countries in the world").

Not being a monster? Eh pear this
Cruel politics,
Who establishes the throne
Of subjects with blood.
No, he wouldn't know at such an undignified price
This heart of mine is committed to the kingdom.[70]

In the central verse—"Is it a merit to us, /Not to be a monster?"—takes to extremes, in the words of Montezuma, that anti-Machiavellian position of the young Frederick which conceives of the ruler not as the absolute sovereign but as a 'first servant' of the state, beholden to his people who entrust themselves to him. This, however, marks a position which—as the further development of the opera also clearly demonstrates—neither allows us to characterize the figure of Montezuma as that of a 'noble savage' nor to equate him, as it were, identificationally with the historical figure of the royal librettist in a kind of camouflage.[71] We are dealing with an intratextual system of references, not infrequently of ironic markings, which—contrary to all the assertions repeatedly found in research[72]—by no means idealizes Montezuma as *bon sauvage* in the eighteenth-century sense,[73] but presents him as a figura, as an aesthetic modelling that is subjected to critical scrutiny in the course of the opera. And this examination, read in the Prussian mirror, will produce a clear result.

The fact that the subject of the Spanish *Conquista* was used on the European stages of the eighteenth century long before the middle of the century and found probably its most famous artistic-musical expression in the field of opera in Antonio Vivaldi's *Montesuma* (1733) has been repeatedly and rightly pointed out. Yet the tradition of the Prussian court opera since 1747/48 consisted precisely in "performing French tragedies transformed predominantly into Italian-language operas", so that *Montezuma*, which was performed in 1755 in the opera house Unter den Linden,

[70] *Montezuma. Tragedia per musica*, pp. 14 and 16. "Ach, erhebe nicht eine schwache Tugend so sehr! / Thue ich ettwas anders, als meine Pflicht? / Ich verlange von dir, Freund! nicht nicht Lobes-Erhebungen, / Sondern Rath: Ist es uns ein Verdienst, / Nicht ein Ungeheuer zu seyn? / Ach, verflucht sey jene grausame Staats-Kunst, / Welche in dem Blute der Unterthanen den Thron gründet. / Nein, für einen so unwürdigen Preiß / Könnte sich mein Hertz niemahls zu dem Besitz / eines Thrones entschliessen." ("Ah, raise not a weak virtue so much! / Do I do anything but my duty? / I ask of thee, friend! not praise exaltations, / But counsel: is it our merit, / Not to be a monster? / Alas, curse that cruel art of state, / Which in the blood of subjects founds the throne. / No, for such an unworthy price / My heart could never decide to possess / A throne" (pp. 15 and 17).

[71] See Detering, Susanne: Friedrich II: Camouflage und Selbstverteidigung im Bild Montezumas, p. 203.

[72] See Henze-Döhring, Sabine: *Friedrich der Große: Musician and Monarch*, p. 87. On the problems of such a view, see also Detering, Susanne: Friedrich II: Camouflage und Selbstverteidigung im Bild Montezumas, p. 199.

[73] See on this term Auerbach, Erich: Figura. In Erich Auerbach: *Gesammelte Aufsätze zur romanischen Philologie.* Ed. by Fritz Schalk and Gustav Konrad. Bern/Munich: Francke 1967, pp. 55–92.

may very well be called a "foreigner",[74] even if it presented itself in conformity with the genre as a 'tragedia per musica'.[75]

This stranger, of course, must have seemed strangely familiar to the audience, since he possessed none of that coarse savagery whose archaeology Hayden White explored in his essay on the 'Forms of Wildness' and[76] saw in a "reappearing in the imagination of Western man—as the Wild Man, as the monster, and as the devil". But it is precisely Montezuma, the educated and sensitive Aztec emperor, who does not appear as a wild man and a monster, while at the same time rejecting the merit of not becoming a monster even under the pressure of a cruel policy. The wild man, whose archetypal image has virtually haunted and continues to haunt Western cultural history, is rather Machiavelli's disciple, Hernán Cortés, who proceeded with cunning and treachery, with murderous lust and massacres, and who, while insulting the Mexica as uneducated barbarians, is always branded as the real barbarian, both in his speeches and in his actions.

From the beginning, Frederick II left no doubt about this constellation, as the letter of October 1753 to Algarotti showed. It is Frederick's tyrannical Cortés—and not his enlightened Montezuma—who shows that "abyss into which mankind might fall",[77] that is, that abyss which is able to open so easily under the thin cover of occidental culture, not only in Prussia.

Thus Montezuma, brought to the Berlin stage, is neither the evil savage nor the noble savage,[78] but the figura of that ruler drawn in the *Antimachiavel*, who is now, of course, *publicly* subjected to an endurance test in the closed space of the Prussian opera as in an experimental room. At the beginning of the opera he still stands there in radiant splendour, knows himself and his country to be supremely happy, and understands his own actions to be in complete accord with the wishes and needs of his subjects, whose will he represents. But with the appearance of Hernán Cortés and his soldiers, who are determined to do anything, this position is put to the test, and even more so, a test whose consequences the ruler—despite all warnings—does not understand at the beginning and whose tragedy he is defencelessly exposed to at the end. The outcome of this test is the collapse of the state and the execution of the

[74] Müller-Lindenberg, Ruth: Power and Senses? On the Concept of the Exhibition, p. 8. On the history of reception from the perspective of Oswald Spengler, see also Birkenmaier, Anke: *Versions of Montezuma. Lateinamerika in der historischen Imagination des 19. Jahrhunderts*, p. 6 f.

[75] On Frederick's traditionalist orientation as a musician as well as a composer in the context of his time as Crown Prince of Prussia, see also Kunisch, Johannes: *Friedrich der Große. Der König und seine Zeit*, p. 85.

[76] White, Hayden: The Forms of Wildness: Archaeology of an Idea. In White, Hayden: *Tropics of Discourse. Essays in Cultural Criticism.* Baltimore/London: The Johns Hopkins University Press 1978, p. 180.

[77] Ibid.

[78] From the immense literature on the *bon sauvage*, only Kohl, Karl-Heinz: *Entzauberter Blick. Das Bild vom Guten Wilden und die Erfahrung der Zivilisation.* Berlin/Frankfurt a. M.: Medusa 1981; and Duviols, Jean-Paul: Contribution des récits de voyage à la formation de l'image du 'Bon Sauvage' Américain'. In: *Les Amériques et l'Europe. Voyage—émigration—exil.* Actes de la 3ème semaine latino-américaine. Toulouse: Université de Toulouse/Le Miral 1985, pp. 27–36.

ruler who was so popular with the people, at least at the beginning. An instructive piece, indeed.

Thus one might venture the thesis that in Friedrich's libretto and in Carl Heinrich Graun's opera, the early Frederician anti-Machiavel is simultaneously staged, tested for his ability to resist, and ultimately found to be unable to survive as a ruler. Montezuma may have been a stranger to the Berlin audience: The figure of the ruler sketched out at the beginning of the opera was not. In this transmedial transfer from the theory of the philosophical-political treatise to the staged practice of the real opera, a fundamental transformation takes place insofar as, in the thunderstorm of *special effects* in the stage and auditorium, the fact is glaringly illuminated that the purity of the figura—as Friedrich noted in his *Antimachiavel*—can be[79] maintained neither in "state[s] life" nor in its significance for "life itself". The ruler as antimachiavel collapses as ruler under the onslaught of Machiavel in the shortest possible time—and with him his entire state system, which in the end goes up in flames and disappears from world history. Such a thing, however, was to be prevented for Prussia.

For Montezuma tries to maintain his, on the one hand abstract, but on the other hand thoroughly tested in the state reality of communal life, conviction of ethically founded good government even in the face of all those unscrupulous enemies whom the sea has, as it were, spat out and thrown into his own country. Their numerical inferiority is of no consequence. Frederick Cortés recognizes very early on the ethics of good rule as a weakness of the system and seeks to counter it with all his might by staging his own power, which relies precisely on the confusion of all the senses. He does this with resounding success.

Thus it is the pistol shots, of which Friedrich reports not by chance in his letter to Margravine Wilhelmine of Bayreuth shortly after the performance, which fired on stage refer to the cunning tactics of the Spaniards, determined to do anything, to confuse first of all the senses of their opponents. Once the senses are overwhelmed, power over the Aztecs can be exercised almost without limit. It is the *special effects* of an inhuman stagecraft that allow the Spaniards to drive the fear-stricken Aztecs before them and bring Montezuma to his fatal fate. He who procures power over the senses of his enemies as well as of his subjects makes of them whatever sense he wishes to make of the history of his own power.

Friedrich's opera *Montezuma* demonstrates this clearly time and again and makes it heard. In full possession of the ruthless certainty with which the Spaniard Narvès makes it clear to Montezuma in the seventh scene of the first act that the whole world has already been conquered, the Spanish troops under the leadership of Hernán Cortés proceed at the end of the fifth scene of the second act to the sound of a warlike symphony with all determination and brutality against the Mexica, who already cease all resistance at the sound of the first, for them strange incomprehensible shots. Not the honour and the noble values, but the resistance of Montezuma is broken, as he appears at the beginning of the immediately following sixth scene as

[79] Frederick the Great: *Antimachiavel*, p. 331. In the original French: Frederick the Great: *Anti-Machiavel ou Examen du Prince de Machiavel*, p. 3: "la politique", "la pratique".

"attonito, e spaventato".[80] The Spaniards, however, know themselves to be the protagonists of a world whose history they have long since begun to rewrite from their own point of view. And yet they are barbarians and rule over a land that they are only capable of plundering and destroying.

This is certainly not the image of a ruler to which Frederick II felt committed. And yet: Had he not, in his *anti-Machiavel*, painted the picture of the noble prince so short a time before his accession to power, broken all the rules of inter-state treaty loyalty and reciprocity after only a few months on the throne and invaded Silesia with cunning and trickery, but above all with brutal force of arms, in order to successfully break the rich province out of the Habsburg Empire, which was temporarily weak in leadership after the death of Emperor Charles VI, and annex it to his own empire? Certainly Frederick did not want to be confused with the figure of the barbarian Cortés; but neither did he want to be a Montezuma whose resistance succumbed to the terrifying gun-smoke and whose empire crumbled under the hoofs of horses unknown to the Mexica. *Montezuma* is a play about power in the medium of power; it therefore makes sense only insofar as it reflects power over the senses.

Thus, on 6 January 1755, the opera *Montezuma* demonstrates, as it were in its very concrete performance practice, how the figura of a ruler who knows himself to be committed solely to abstract values must perish in the whirlwind of attacks by his enemies and consequently has to summon up all the powers of the media to maintain and, where possible, expand his own power. The opera itself, "l'opéra même", therefore[81] becomes for Frederick II that rousing spectacle, that—as Algarotti wrote to him even before the premiere of *Montezuma*—"spectacle magnifique du Nouveau monde",[82] in which the Prussian monarch and musician can confuse and consequently overwhelm the senses of his subjects not only with words and with music, but also with the exotic colorfulness of the costumes or with the gun smoke of his fireworks. This is the sense, these are the senses that define the New World of Frederick the Great.

Contemporaries already knew it: The Prussian king was not only active as a librettist. He not only exerted the strongest influence on the music as well as (through his frequent presence at rehearsals) on its execution, but also designed stage decorations and costumes, in other words, all the things in opera that are designed for effect and interaction. Since he was wont to intervene in everything like a prince, and even more like a general, the English music historian Charles Burney, after a visit in 1773, probably rightly said of the Prussian monarch: "The king keeps as much manliness in the opera house as in his camp."[83] The horrendous costs of the Linden Opera House were only in apparent contradiction to the king's much-vaunted "austerity policy"; for "power was not only expressed in a powerful

[80] *Montezuma. Tragedia per musica*, p. 66.

[81] Friedrich II: *Correspondence of Frederick the Second, King of Prussia, with Count Algarotti*, p. 98.

[82] Algarotti's letter of 11 January 1754 from Venice to Frederick II, in: Frederick II: *Correspondance de Frédéric Second, Roi de Prusse, avec le Comte Algarotti*, p. 100.

[83] Citation after Leithold, Norbert: *Friedrich II. von Preußen. A cultural-historical and illustrated panorama from A to Z.* Frankfurt a. M.: Eichborn (Die Andere Bibliothek) 2011, p. 327.

army"—the opera was an excellent "instrument of power", of[84] which Frederick knew how to make excellent use.

Even today, the figurines of the Mexican emperor[85] sculpted by Christian Gottlob Fechhelm bear witness to the enormous splendor of an outfit that the Prussian monarch undertook at enormous material expense and with enormous fees for his opera stars[86] as well as for his orchestra, his decorators and stage architects.[87] With its power over the senses of its subjects, opera in its sensually perceptible fullness itself demonstrated the meaning of power, embodied in the royal librettist. It is a spectacle of power that does not celebrate power as a spectacle, but rather the meaning of power in the power over the senses.

Can Frederick's princely libretto and the performance of 'his' opera in the music of Carl Heinrich Graun, in the theatrical architecture of Giuseppe Galli Bibiena, in the arias of Giovanna Astrua and hundreds of other participants be understood as a prelude to that Berlin debate about the New World which, from the end of the sixties of the eighteenth century onwards, heated tempers in a *République des Lettres* that had long since become worldwide?

The answer to this question cannot be simple, but will have to be differentiated. For Frederick II, quite contrary to the Prussian traditions of court opera that he himself protected, did move the setting to the spatially and temporally distant land of the Aztecs, but he did not do so in order to set in motion an anthropological dispute of any kind about the inhabitants of both the Old and the New World.[88] It is true that he unfolded a critical, even scathing review of the protagonists of the first phase of accelerated globalization from the experience of a phase of accelerated globalization in which the new powers of European expansion loudly proclaimed their own superiority over Spain and Portugal on both the material and—thanks to the aid of the *leyenda negra*—on the ethical-moral level. But he was certainly not concerned with the programmatic revaluation of non-European cultures, as we see developing

[84] Leithold, Norbert: *Friedrich II. von Preußen. Ein kulturgeschichtliches und bebildertes Panorama von A bis Z*, p. 327.

[85] See Henze-Döhring, Sabine: *Friedrich der Große*, p. 89.

[86] See on this Leithold, Norbert: *Friedrich II. von Preußen. Ein kulturgeschichtliches und bebildertes Panorama von A bis Z*, p. 327; there you will also find more detailed information, for example, for the prima donna Astrua or the castrato Salimbeni.

[87] Ibid., p. 328 as well as Müller-Lindenberg, Ruth (ed.): "Macht und Sinne? Zum Konzept der Ausstellung Friedrichs Montezuma." In: Müller-Lindenberg, Ruth (ed.): "*Friedrichs Montezuma*". *Power and Senses in the Prussian Court Opera*. Publikation zur Sonderausstellung im Musikinstrumenten-Museum, Berlin 2012, pp. 9–18.

[88] That Frederick II was only concerned with myths and legends surrounding the conquest of Mexico is a recurring claim in research; see, for example Detering, Susanne: Friedrich II: Camouflage und Selbstverteidigung im Bild Montezumas, pp. 198; in fact, the Prussian monarch demonstrably resorted to the *Historia* de la conquista de México by Antonio de Solís y Ribadeneyra from 1684, which is considered a standard historiographical work; see on this, among others, Klüppelholz, Heinz: Die Eroberung Mexikos aus preußischer Sicht—Zum Libretto der Oper 'Montezuma' von Friedrich dem Großen, pp. 68–79.

on an anthropological basis in Prussia half a century later in the work of Alexander von Humboldt,[89] but rather with the creation of an aesthetic testing ground in which the historically found as well as the artistically invented should be brought to a directly sensual experience. Opera as a political demonstration of power as well as a demonstration of political power provided and created the projection space and the sounding board with the help of which the question of power could be posed on a global scale. On the level of Prussia as well as on that of the entire world.

At the latest since the liquidation of the remnants of its overseas trading company, Prussia was no longer a protagonist within the phase of accelerated globalization that began again around the middle of the eighteenth century. Perhaps it was precisely this distance that created the opportunity to analyse and theoretically assess the globalisation thrusts all the better. Thus it is certainly no coincidence that writers and philosophers familiar with non-European issues, such as Maupertuis or Voltaire, such as de Pauw, Pernety or Raynal, frequented Potsdam and Berlin. Certainly the Berlin debate played out in the provinces. But at the Prussian court, as in the emerging Berlin, a preoccupation with non-European issues developed with a certain continuity. This raised fundamental problems of the phases of accelerated globalization, the first theorist of which was probably a researcher, scholar and writer born in provincial Berlin: The younger of the two Humboldt brothers.

If Alexander von Humboldt's writings can be understood as a late, but by no means belated, response to the second phase of accelerated globalization, then the works of Cornelius de Pauw or Antoine-Joseph Pernety, which were created in Potsdam and Berlin, but also the opera design of Frederick's *Montezuma*, are complex and contradictory reactions to a changing political, economic and cultural situation, which we may understand as the second phase of accelerated globalization. As an opera by Frederick II, *Montezuma* is certainly not part of the Berlin debate that flared up at 'his' Academy; but at the same time it is considerably more than its mere prelude.

Friedrich's 'Tragedia per musica' poses the question of power in its global dimension, a question that can only be re-examined today under changed conditions, whereby the Prussian monarch allows the confusion of the senses, which Hernán Cortés once used with horse stamping and artillery to overcome the Mexica, to be experienced sensually as a power factor. Today, of course, other media systems have long since taken the place of opera in this function. But the overwhelming of the senses staged by Friedrich demonstrates not only the power that the senses exercise over our thinking and our lives, but also the necessity of always confronting a violence that intoxicates the senses with the critical question of the meaning of power, if a new world is to be aesthetically preconceived and prepared. But let us let

[89] See on this anthropological dimension Ette, Ottmar (ed.): *Das Buch der Begegnungen: Menschen – Kulturen – Geschichten aus den Amerikanischen Reisetagebüchern*. Munich: Manesse 2018, for a juxtaposition of De Pauw and Alexander von Humboldt see ders: "Not just brought about by chance": reflections on globalisation in Cornelius de Pauw and Alexander von Humboldt. In: *Studies in Travel Writing* (Nottingham) XV, 1 (February 2011), pp. 3–25.

Eupaforice, Montezuma's bride, that fearless indigenous woman who, unlike all the other characters, is a pure invention of Frederick's, have the last word, as at the end of the second act:

> According to Heaven the bold,
> Who with courage armed
> Who despises the rigor of Fate,
> And does not know how to fear.[90]

[90] *Montezuma. Tragedia per musica*, p. 102. "Heaven stands by the bold, / If he arms himself with courage, / Not frightened, and despises the severity of / Fate" (p. 103).

Dom Pernety and the Berlin Debate

Travel/Writing

In one of the most important comments on his *Discours sur l'origine et les fondements de l'inégalité parmi les hommes*, Jean-Jacques Rousseau critically examined the state of anthropological knowledge in Europe at the time in 1755. In doing so, he emphasized the asymmetry, which could not be overlooked by most European philosophers, between the state of knowledge of travelers from the most diverse European countries on the one hand and the state of reflection of all those philosophers on the other, who made and developed their generalizing reflections and theories on the diversity of the human race not least on the basis of more or often less extensive and attentive readings of travel reports.

In this context, Jean-Jacques Rousseau stressed the indispensable need to break this fundamental asymmetry within the worldwide circulation of knowledge, which, moreover, was not least aggravated by the lack of preparation of the vast majority of travelers. At the same time, however, he also noted a certain lack of (scientific) orientation on the part of some *hommes éclairés*, who had nevertheless exposed themselves to multiple dangers and undertaken long arduous journeys:

> The Academicians who travelled through the northern parts of Europe and the southern parts of America were more interested in visiting them as Geometers than as Philosophers. However, as they were both, one cannot consider as completely unknown the regions which were seen and described by the La Condamine and the Maupertuis.[1]

The author of the *Discours sur l'inégalité*, as this *Second Discours* is also often called in abbreviated form, left no room for doubt: It was not from the geometers but the philosophers that Rousseau hoped for decisive improvements in the state of

[1] Rousseau, Jean-Jacques: *Discours sur l'origine et les fondements de l'inégalité parmi les hommes*. In: Jean-Jacques Rousseau: *Œuvres complètes*. Vol. III. Edition published under the direction of Bernard Gagnebin and Marcel Raymond with, for this volume, the collaboration of François Bouchardy, Jean-Daniel Candaux, Robert Derathé, Jean Fabre, Jean Starobinski and Sven Stelling-Michaud. Paris: Gallimard 1975, p. 213.

O. Ette, *Mobile Prussia*, https://doi.org/10.1007/978-3-476-05839-3_3

knowledge about the non-European world. Although the citizens of Geneva readily acknowledged the content and quality of some of the travelogues published in the eighteenth century, he did not hide his fundamental criticism of the general level of anthropological or ethnological knowledge, whereby, for all his admiration for the work of a Buffon, he did not exclude from this criticism the field of natural history and, in particular, the sources used by its representatives. After mentioning a few less credible accounts, he stated with all desirable clarity:

> Apart from these relations, we do not know the peoples of the East Indies, frequented only by Europeans more interested in filling their purses than their heads. The whole earth is covered with Nations of which we know only the names, and we interfere to judge the human race! Let us suppose a Montesquieu, a Buffon, a Diderot, a Duclos, a d'Alembert, a Condillac, or men of that calibre travelling to instruct their compatriots, observing and describing as they know how, Turkey, Egypt, Barbary, the Empire of Morocco, Guinea, the countries of the Caffres, the interior of Africa and its eastern coasts [... Let us suppose that these new Hercules, on their return from these memorable journeys, would then write at leisure the Natural, Moral and Political History of what they had seen, we would see a new world emerge from under their pen, and we would thus learn to know our own.[2]

Due to their all-dominating personal as well as commercial interests, most European travelers, with their reports so numerous but often so unsubstantiated and, above all, so guided by interests, do not escape Rousseau's ultimately damning judgement. Yet the author of the *Discours sur l'inégalité* in no way denied the fundamental importance of travel for the expansion of human knowledge and the extensive circulation of knowledge in the European-extra-European field of tension. Quite the contrary: For him, it was a matter of sending out well-prepared travelers, 'philosophers' (in the eighteenth-century sense), who had not only a knowledge (*savoir*), but even more a *savoir faire* and a *savoir voir*, and who had to be able, on their return to Europe, to pass on what they had seen to their compatriots as well as to their reading public within a *République des Lettres* of the Enlightenment that tended to span the globe. For only in this way was it possible to create new foundations for new knowledge.

Rousseau's ideas are of the greatest epistemological scope: *Savoir faire* and *savoir voir* were thus to turn for the traveller's future readers into a *savoir faire voir*, knowledge and technique of foresight, which could not only modify the European view of the New World and transform the *nouveau monde* into a *monde nouveau* but also fundamentally change the view of European countries themselves.[3] Foreign knowledge always implies self-knowledge, and even more: Self-knowledge cannot be had without foreign knowledge. Moreover, it cannot be overlooked that within a worldwide relationality—though undoubtedly dominated by Europe—the knowledge of travelling European philosophers had to be of use to the inhabitants of other

---

[2] Ibid., p. 213.

[3] I first presented these considerations from a different perspective and with a view to the epistemology of eye and ear in Ette, Ottmar: Diderot et Raynal: l'oeil, l'oreille et le lieu de l'écriture dans l'Histoire des deux Indes'. In: Lüsebrink, Hans-Jürgen/Strugnell, Anthony (eds.): *L'Histoire des deux Indes': réécriture et polygraphie*. Oxford: Voltaire Foundation 1996, pp. 385–407.

areas and regions of the world as well. Rousseau's arguments are indebted to the ethics with universalizing claims on which the (European) *République des Lettres* was based.

In the passage cited, seeing and writing, in their connection with travel, appear as complementary acts that, in their sequence, produce a sense (concerning) the New World, which Rousseau would then also be prepared to believe: "you have to believe them."[4] But this belief, for Rousseau, is not based on writing, *écriture*, alone. It is not the movement of discourse, but the movement of travel that grants a direct view of things, that change of place, in other words, that makes possible an immediate seeing of the other, that gives authority to writing about the other, and thus the only authorship in the strong sense. In these passages of the *Second Discours*, the credibility of this authorship is consequently tied back to the legitimation through an eye-witnessing which, with regard to the knowledge of distant countries, presupposes one's travel.

An idea glows here that is of the greatest topicality and urgency for our time, for our own scientific and science-political epistemologies (and 'self-evident truths'). For a long time—and right up to the present day—the idea persisted that one had to clearly separate the 'general', the generalizing theory from the so-called regional sciences, the *area studies*, as a 'systematic' or systematizing science, since the latter would only be able to produce a regionally limited knowledge—for instance about Turkey, North Africa or South America, to stay with the examples of Rousseau. Nothing, however, is more presumptuous and inadequate to the objects—and this, too, will be shown by the Berlin debate on the New World in historical retrospect—than such a separation.

The allegedly 'general' theory is mostly based on extremely incomplete knowledge, which is usually limited to the knowledge of a more or less small-scale and, at best, European section, which is unaffectedly and unquestioningly set as normative. However, in order to be able to adequately substantiate and develop a general theory, knowledge of the most diverse areas is indispensable. Only then can Rousseau's goal of grasping our world in its entirety in a new way take effect: "we ourselves would see a new world emerge from under their pen, and we would thus learn to know our own."[5]

One could juxtapose Rousseau's extensively quoted reflections with a passage from the pen of Diderot, whom he mentions in the above quotation, first published in[6] 1780 in the eleventh book of the third edition of the *Histoire philosophique et politique des établissements et du commerce des européens dans les deux Indes*. In a sense, it is a response and, even more, an astute epistemological replica twenty-five years after the publication of the second *Discours*. At the beginning of Rousseau's Reflections, we find his complaint about never having found 'deux

[4] Rousseau, Jean-Jacques: *Discours sur l'origine et les fondements de l'inégalité parmi les hommes*, p. 214.

[5] Ibid, p. 213.

[6] See Duchet, Michèle: *Diderot et l'Histoire des Deux Indes ou l'Ecriture Fragmentaire*. Paris: Nizet 1978, p. 84.

hommes bien unis, riches, l'un en argent, l'autre en génie, tous deux aimant la gloire', fame for which one would be willing to sacrifice 'twenty thousand thalers', the other 'ten years of his life' in favour of a successful journey around the world.[7]

We are not interested here in the remarkable relationship between the figures, but in the fact that Diderot resumed the discursive structure of this passage, at the same time giving a new twist to its critical sense:

> The rich man sleeps; the scientist watches, but he is poor. His discoveries are too indifferent to governments for him to be able to solicit help or hope for rewards. There is more than one Aristotle among us; but where is the monarch who will say to him: My power is at your command [...].[8]

A quarter of a century after the publication of Rousseau's second *Discours*, the link between the rich man and the genius, between the monarch and the scholar, has not become more realistic. Unlike Rousseau, who would have loved to send a Buffon, as well as a Diderot, on a world tour, the co-author of Guillaume-Thomas Raynal's *Histoire des deux Indes* was in no way inclined to leave his study in Paris and sail around the world. And in a passage from his pen, he cited epistemological rather than personal reasons for doing so:

> The contemplative man is sedentary; & the traveller is ignorant or a liar. He who has received genius as a share disdains the minute details of experience; & the maker of experiments is almost always without genius.[9]

For Rousseau, knowledge is based on seeing, *sa-voir* on *voir*. If the link between the man blessed with money and the man blessed with genius is accidental, the unity between the philosopher and the traveller, between *philosophe* and *voyageur*, is for Rousseau conscious and conceptual: It is programmatic. We do find in Diderot the division between the mighty and the learned, but this *savant does* not undertake any physical journeys, his movements being purely intellectual. Diderot assigns him a place which he will not leave: The place of work and the place of writing are not different from each other in a purely spatial sense. Diderot contrasts *homo contemplativus* not with *homo faber* but with the traveller, a kind of *homo migrans*, who suffers either from a lack of knowledge (*ignorant*) or from a lack of truth and veracity (*menteur*) and who must, therefore, for Diderot, be downgraded to secondary importance at best in a system of hierarchized circulation of knowledge. *A beau mentir qui vient de loin*—and from today's perspective, one could almost believe that this proverb was also coined for those sciences that were once somewhat unhappily referred to as 'remote sciences' a few years ago.

---

[7] Ibid., p. 213.

[8] Raynal, Guillaume-Thomas: *Histoire philosophique et politique des établissements et du commerce des européens dans les deux Indes.* Tome cinquième. Genève: Chez Jean-Léonard Pellet, Imprimeur de la Ville & de l'Académie 1781, p. 43.

[9] Ibid.

But let us stay in the eighteenth century. It was not for nothing that Denis Diderot had the dialogue partners of his *Supplément au voyage de Bougainville* debate whether the famous French explorer had not rather been a sedentary inhabitant on the planks of his floating house ("maison flottante")[10] during his circumnavigation of the world, while the reader of Bougainville's famous travelogue must be regarded as the actual world traveller since it was he who—apparently immobile on the floorboards of his fixed house—had circumnavigated the globe by virtue of his reading.[11] One could certainly speak here with good reasons not of a Diderotian *paradoxe sur le comédien*, but of the *paradox* about the traveler, of the *paradoxe sur le voyageur*, who in his highest form is the reader—and if he were a reader of the book of the world.[12]

Without being able to elaborate here on the fact that the epistemologically relevant distinction between the travelers and those who stayed at home was by no means an invention of the eighteenth century, but rather ran through the entire Western discourse on the New World since the first phase of accelerated globalization. Without being able to elaborate here on the fact that the epistemologically relevant distinction between travelers and those who stayed at home was by no means an invention of the eighteenth century, but rather ran through the entire Western discourse on the New World since the first phase of accelerated globalization, and was incidentally already present in antiquity with regard to the interplay of eye and ear as sources of information about an unknown world,[13] it should nevertheless be pointed out that beyond the epistemological contrast between Rousseau and Diderot that is marked here and is indeed striking, the two great philosophers of the European Enlightenment agree with each other on a not insignificant point. Rousseau's travelling philosopher, the *philosophe voyageur* chooses the same place as Diderot's sedentary *homo contemplativus* to write down his work: The European desk.

For both European *philosophes*, the hermeneutic movement pattern of the journey as a process of understanding is always based on a circle. And so it is by no means coincidental that Rousseau speaks of a journey around the world in the passage quoted above. Once the travelling philosopher has returned home and has been transformed into a 'new Hercules' because of his toil and suffering, he will soon sit down at his desk and write down his views. The fact that the writing down of the text, the actual production of the text, 'of course' does not take place overseas, but in Europe, seems almost natural to both Europeans.

---

[10] Diderot, Denis: Supplément au Voyage de Bougainville ou Dialogue entre A et B. In: Denis Diderot: *Œuvres*. Edition établie et annotée par André Billy. Paris: Gallimard 1951, p. 964.

[11] See Ette, Ottmar: 'Le tour de l'univers sur notre parquet': lecteurs et lectures dans l'Histoire des deux Indes'. In: Bancarel, Gilles/Goggi, Gianluigi (eds.): *Raynal, de la polémique à l'histoire.* Oxford: Voltaire Foundation 2000, pp. 255–272.

[12] See Blumenberg, Hans: *Die Lesbarkeit der Welt.* Frankfurt a. M.: Suhrkamp 1986.

[13] See on these connections more comprehensively the third chapter in Ette, Ottmar: *Literatur in Bewegung. Raum und Dynamik grenzüberschreitenden Schreibens in Europa und Amerika.* Weilerswist: Velbrück Wissenschaft 2001.

The place of writing, as a place of knowledge production within a profoundly asymmetrical circulation system of knowledge, can only be located in Europe. Reading in the book of nature may or must take place worldwide; the writing down of this reading of nature, however—neither a Rousseau nor a Diderot needs to address this—can 'naturally' only take place in the European centres of knowledge. The whole world may be a realm of signs: But Europe sees itself as the realm of knowledge.

Old World/Other World

The *Berlin debate about the New World*, which must be regarded as an eminently important part of what Antonello Gerbi called the 'Dispute about the New World'[14] in a study that has long since become canonical, is also situated within the problematic relationship between travel and knowledge and travel and writing outlined here by Jean-Jacques Rousseau and Denis Diderot. One of the two most important protagonists of this Berlin debate, which was registered and commented on with the greatest attention not only in Europe but also overseas, Antoine-Joseph Pernety, had once travelled in the function of a ship's chaplain with none other than Bougainville, not around the entire globe but to what Amerigo Vespucci once called the 'New World'. This fact was to play an important role in those disputes that began in the ambitious Prussian king Frederick II's up-and-coming capital—who himself provided a kind of sensual, power-political 'prelude' to this debate with his libretto for the opera *Montezuma*, which was first performed at the Linden Opera on 6 January 1755[15]—but quickly met with a worldwide response. Prussia may not have been on the map in colonialism and the colonialist division of the earth, but it was very much on *the map* in the colonial debate.

With some of the central questions of Enlightenment anthropology, but also with the problems of epistemology of perception theory presented here the Benedictine Antoine-Joseph Pernety, formerly persecuted by the Inquisition in Avignon and having fled to Prussia, already dealt in his *Journal historique* published in 1769 in French in Berlin, in which the on the title page proudly called "Membre de l'Académie Royale des Sciences & Belles-Lettres de Prusse" ("Member of the Royal Prussian Academy of Science and Literature") and "Bibliothécaire de Sa Majesté le Roy de Prusse" ("Librarian of His Majesty the King of Prussia") reports on his journey under the direction of Louis-Antoine de Bougainville to the "Bibliothécaire de Sa Majesté le Roy de Prusse" ("Librarian of His Majesty the King of Prussia"). "Belles-Lettres de Prusse" ("Prussian Belles-Lettres") and as "Bibliothécaire de Sa Majesté le Roy de Prusse" ("Librarian of His Majesty the King of Prussia") reported of his voyage under the guidance of Louis-Antoine de

[14] See Gerbi, Antonello: *La disputa del nuovo mondo. Storia di una polemica: 1750–1900*. New edition edited by Sandro Gerbi. With a profile of the author by Piero Treves. Milano/Napoli: Riccardo Ricciardi editore 1983.

[15] See the second chapter in this volume.

Bougainville to the tropical coasts of present-day Brazil, to the Malvinas Islands, and to the southern tip of the American continent.[16]

In his Journal Pernety, again and again, emphasized the fundamental otherness of the New World; thus he also described on no less than twenty pages the crossing of the equator on November 10, 1763, and thus that scenery which took place in such a way or comparable form on all French as well as European ships passing the equatorial line, the central area of the tropical belt, and entering another hemisphere. For a long time, the tropics had become the planetary space of movement *par excellence* for the European seafaring nations,[17] but at the same time, they represented the threshold area of a transition which, in the case of this ship heading south to Brazil, may be described as a symbolic transition from the eastern to the western hemisphere, but at the same time also from the northern to the southern hemisphere. This double transition is marked by the Frenchman's voyage report in a quite impressive way.

Dom Pernety's detailed account of the "*Baptême de la ligne*,"[18] which still in the blazing midday heat on the deck of the French ship with the setting up of a basin with seawater, a throne for the "Seigneur Président de la Ligne" ("Lord President of the Line") as well as further seats for the "Chancelier" ("Chancellor") as well as the (from the point of view of the *aumônier* certainly especially interesting) "Vicaire" ("Vicar") together with an assembly of the entire crew and all passengers,[19] emphasizes in detail and not without literary skill the ceremonial and ritual character of this baptism of the equator. The carnivalesque features of the scene are unmistakable, as first of all—with Captain Bougainville's consent, of course—a sailor, seconded by six ship's boys and dressed and painted in sheepskin, enters the stage, his disguise and transformation precisely described,[20] before the *bon-homme de la Ligne* himself, also dressed in sheepskin, takes command on board, surrounded by his decked-out court. The spectacle of travestied alterity begins.

The crossing of the equatorial line, comparable to the time-space of the carnival, turns the 'natural' hierarchies upside down: A *monde à l'envers* that seems to combine the carnivalesque element with the idea of the antipodes—a world whose inhabitants are, as it were, upside down and in which the trees grow downwards. The reading public is, thus, vividly made aware that it is following the traveller into another world, an alien world, which has its store of signs. It is, therefore, worth

[16] See Pernety, Antoine-Joseph: *Journal historique d'un Voyage aux Iles Malouïnes en 1763 & 1764, pour les reconnoître, & y former un établissement; et de deux Voyages au Détroit de Magellan, avec une Relation sur les Patagons.* 2 vols. Berlin: Etienne de Bourdeaux 1769.

[17] See on this Ette, Ottmar: Discurse der Tropen—Tropen der Diskurse: Transarealer Raum und literarische Bewegungen zwischen den Wendekreisen. In: Hallet, Wolfgang/Neumann, Birgit (eds.): *Raum und Bewegung in der Literatur. Die Literaturwissenschaften und der Spatial Turn.* Bielefeld: transcript 2009, pp. 139–165.

[18] Pernety, Antoine-Joseph: *Journal historique d'un Voyage aux Iles Malouïnes en 1763 & 1764*, vol. 1, p. 95.

[19] Ibid., p. 96.

[20] Ibid., p. 98 f.

paying some attention to the literary elaboration of this scene, which undoubtedly marks the travel-literary location of a high point[21] in Pernety's *Journal historique.*

Without being able to reproduce the literary scene designed by Pernety in all its details, it is nevertheless instructive to observe that the staging of a counter-order, which symbolically comes into effect for one day between the tropics, presents those twists and turns that represent the tropics as a *different* world *from* Europe under the sign of the dividing line of the equator. The extent to which this other order is projected onto the 'other' sex from a patriarchal perspective that is not itself subverted may be demonstrated by the following quotation from the captain's fruitful 'promises':

> Things being thus arranged, the Vicar said to Mr. de Bougainville: "To be received into the noble & powerful society of the Lord President de la Ligne, you must first make some commitments, which you promise to observe. These commitments will only concern justifiable matters. Mr. de Bougainville replied... Do you promise," said the Vicar, "to be a good Citizen, & for this purpose to work for the population, & not to let the girls idle, whenever the favourable occasion arises? Ä I promise."[22]

The series of promises, which is concluded with an oath and a ritualized baptismal scene in which, from a Christian point of view, the symbolism of the beginning of another life is always present, opens a sequence of further merrymaking, in which two paradisiacal "Demoiselles Acadiennes"[23] as well as all kinds of teasing games, which subsequently take place on Bougainville's frigate under the command of the *bon-homme de la Ligne*, may not be missing.

The projection of male-dominated sexual permissiveness onto the tropics, onto the 'other' world, is part of the core of European images of distant lands. Thus, it is no coincidence that discussions about the sexuality and sexual life of indigenous populations, influenced by European sexual fantasies, will also play an important role (although not pursued further here) in the Berlin debate about the New World. However, the patriarchal-colonialist trait of the European discourse is unmistakable.

Within the narrative sequence of the *Journal historique*, the detailed account of the baptism of the equator sharpens the reading public's attention to the discursive (and epistemologically fundamental) otherness of the 'other' world. For between Cape Verde and the Brazilian coast, the frigate not only crosses the equatorial line but symbolically enters with its crew and passengers into another world, into a new life for the French crew. What is decisive is not that the forms of baptism differ from nation to nation, but also from ship to ship; what is decisive is rather the fact that this *baptême* is always performed on board and represents a vested right of the crews.[24] Thereby the baptism is not primarily bound to the crossing of the equatorial line itself, but the entry into the zone of the tropics, as Pernety explains in conclusion:

---

[21] See on this concept of travel literature theory Ette, Ottmar: *Literatur in Bewegung*, pp. 53–58.

[22] Pernety, Antoine-Joseph: *Journal historique d'un Voyage aux Iles Malouïnes en 1763 & 1764*, vol. 1, pp. 107 f.

[23] Ibid., p. 109.

[24] Ibid., p. 111.

> When the Ship in its course is not to pass the Line, but only the Tropic, those of the Crews who have already passed it, not wishing to lose this right of tribute, have taken it into their heads to name the Tropic, the eldest son of the good man [of] the Line, as the presumptive heir of his rights. Consequently, when the Tropic passed, they played the same joke as the others under the Equator.[25]

Thus, the tropics as a global space of movement represent the 'actual' dynamic core of European ideas of alterity. The fact that this scenery, which also carnivalized religious elements, was subsequently transferred to other 'borderlines' and—as the ritual 'polar baptism' on board the German research ship 'Polarstern' can still show today—is also celebrated onboard many ships in our days, does not in any way diminish the symbolic significance of the figure of movement of the tropics, which becomes recognizable based on these *rites de passage*. For these Western-designed tropes are tropes of transformation, of metamorphosis, in which the transfer from an 'Old' to a 'New' world is accompanied by a fundamental transformation of all customs, habits and living conditions. The New World is another world, represented by the Old World as a clearly divorced counter-world.

Yet this otherness of the tropics and tropical America is all too often signified in European scholarly discourse by fundamental inferiority. However, much the historical contexts may have changed in the second half of the eighteenth century, during the second phase of accelerated globalization, and the Iberian powers have long since been replaced by France and England as globalizing leading nations: Those tropes that determine the view of the tropics from a European perspective can be found largely unchanged in the writings of the *philosophes of* the *European* Enlightenment.

Without ever being heard in Europe for a long period, American Enlightenment thinkers, such as the New Spanish Jesuit Francisco Javier Clavijero[26] vehemently and with good reasons resisted being labelled the inferior other of Europe by contemporary European scholarship in the wake of Buffon. Beyond its fullness, however, the tropical world of the Americas appears in the eyes of most European philosophers as a trap that configures an 'other' world fundamentally divorced from their 'own', entirely in the sign of its asserted fundamental inferiority.

Thus, the description of the crossing of the equator on November 10, 1763, in Antoine-Joseph Pernety's *Journal historique* already points in advance to fundamental disputes that were to be conducted in the course of the Berlin debate about the New World that broke out a few years after this voyage. For it was precisely the thesis of the fundamental weakness and inferiority of the New World that had generated images in European Enlightenment literature, which in the *Recherches philosophiques sur les Américains*[27] by Cornelius de Pauw, published in Berlin in 1768

[25] Ibid., p. 112.

[26] See the book, first published in Italian in exile, by Clavijero, Francisco Javier: *Storia Antica del Messico*. 4 vols. Cesena: Gregorio Biasani 1780.

[27] See Pauw, Cornelius de: *Recherches philosophiques sur les Américains, ou Mémoires intéressants pour servir à l'Histoire de l'Espèce humaine*. 2 vols. Berlin: Chez Georges Jacques Decker, Imp. du Roi 1768–1769.

and 1769, reached the level of sometimes almost apocalyptic imagery under the sign of the degeneration thesis.

For de Pauw, a cleric born in Amsterdam in 1739 and who died in Xanten in 1799, it was finally evident that starting from the tropics, diseases and epidemics, such as syphilis or yellow fever would spread across the entire globe and had to endanger the continuity of human life on the planet. He had understood that his time was the time of a considerably intensified exchange, a time in which that process *de longue durée* which we now call 'the' globalization had picked up considerable speed again.

Let us not prejudge the specific sequence of events surrounding the Berlin debate on the New World. But Antoine-Joseph Pernety in his speech of September 7, 1769, before that Berlin *Académie des Sciences & Belles-Lettres* of which he was a member, Antoine-Joseph Pernety may have developed a decided counter-position against Cornelius de Pauw, but it nevertheless became apparent that in this 'Berlin Debate' on the New World,[28] which was perceived far beyond the borders of Prussia and Europe, de Pauw's position prevailed, and with it a position in which the 'New World', which was also geologically younger than the Old World and consequently had only risen from the waters much later, was stigmatized as a place of fundamental inferiority that existed from the beginning. This as radical as (in Buffon's tradition) popular thesis came, mind you, from an author who at no time had ever left Europe or even set foot in the hemisphere of America he depicted.

Pernety/de Pauw

Right at the beginning of his Dissertation *sur l'Amérique et les Américains, contre les Recherches philosophiques de Mr. de P****,[29] presented to the Berlin Academy *on* September 7, 1769, and published the following year, Antoine-Joseph Pernety built up a front against de Pauw's theses, the strategy of which we can easily recognize from its very beginning:

> Monsieur de P. has just brought to light a work under this title, Recherches Philosophiques sur les Américains. He endeavours in it to give the most unfavourable idea of the new World & of its inhabitants. The affirmative & decisive tone with which he proposes & resolves his questions; the tone of assurance with which he speaks of the soil & productions of America, of its temperature, of the bodily & spiritual constitution of its inhabitants, of their morals & customs, finally of the animals; could lead one to believe that he has travelled in all the countries of this vast expanse of the earth; that he has lived [sic!] One would be tempted to suspect that, among the travellers who have spent long periods there, some have told us fables, have disguised the truth through imbecility, or have violated it through malice.[30]

[28] See on the Berlin debate also the beginning of the preceding chapter.

[29] See also Gerbi, Antonello: *La Disputa del Nuovo Mondo*, pp. 120–125.

[30] I quote from the version included by Cornelius de Pauw in the third volume of his three-volume edition of the *Recherches philosophiques* published in 1770; see Pernety, Antoine-Jseph: Dissertation sur l'Amérique et les Naturels de cette partie du Monde. In: Pauw, Cornelius de: *Recherches philosophiques sur les Américains, ou Mémoires intéressants pour servir à l'Histoire*

Dom Pernety, who—as already described—in the same year 1769 also in Berlin and French presented his two-volume travelogue, played here from the beginning the card of the one who as eye-witness travelled those countries which his adversary Cornelius de Pauw despite the keynote of greatest self-assurance with which he indeed subjected not least also the reports of travelers to a sweeping criticism, never got to see himself. In contrast, Pernety himself, who as a member of the Academy could rightly hope for a positive reception of his reflections by the Berliner Akademie (Berlin Academy), placed himself in the context of that demand for a *philosophe voyageur*, which Rousseau had raised in his *Discours sur l'origine et les fondements de l'inégalité parmi les hommes.*

Therefore, he already emphasized on the first lines of his 'Préface' that he had not approached the world of America solely as a reader: "I had read & reread many reports of America; I had seen with my own eyes most of the things, which are reported there."[31] And it had astonished him greatly to *see* all the things he had seen with his own eyes discarded or disguised: "see contradicted or misrepresented by Mr. de P."[32] As with Rousseau, Pernety's seeing, and thus the traveling philosopher's sense of sight, is crucial to sound judgment.

In other parts of the *dissertation*, too, it is easy to see that Rousseau's world of thought plays a role in Pernety's arguments that should not be underestimated. Thus, Pernety reverses de Pauw's image of the weak, mentally and physically inferior and unmanly (because beardless) Indians into an exactly opposite picture, since the indigenous population possessed "a firm, vigorous health, a life that usually passes the limits of our own."[33] But the strong, long-lived man of nature also possesses an alert mind: "a sound mind, educated, enlightened & guided by a truly natural philosophy, & not subordinated like ours, to the prejudices of education; a noble, courageous soul, a generous, obliging heart: what more does Mr. de P. need to be truly men?"[34]

Pernety cleverly uses the discussion of the human-animal antagonism to position his counter-thesis of the superiority of the *homme naturel*, i.e. the indigenous population of America, against de Pauw, in addition to the frequently repeated mention of American advanced civilizations, such as the Incas, which was inserted early in his text. The representatives of a "truly natural philosophy" should under no circumstances be called savages, "*sauvages*", rather the Europeans would have to put up with such a label: "since our actions are indeed contrary to humanity, or at least to the wisdom that should be the guide of men, who claim to be more enlightened than themselves."[35]

---

de l'Espèce humaine par Mr. de P. Nouvelle Edition, augmentée d'une Dissertation critique par Dom Pernety; & de la Défense de l'Auteur des Recherches contre cette Dissertation. 3 vols. Vol. 3. Berlin: Decker 1770, p. 7 f.

[31] Ibid., p. 4 f.

[32] Ibid., p. 5.

[33] Ibid., p. 114.

[34] Ibid., p. 114 f.

[35] Ibid., p. 115.

If here Pernety mocks the "Beautiful lesson dictated by the lights of pure reason"[36] put forward by his opponent in the Berlin debate, he deplores in the same breath the unhappiness and restlessness of a part of the human race that believes itself to be superior to all other parts. The society of those who are called 'savages', on the other hand, is a community founded on a *contrat social*, on a social contract entirely in the sense of Rousseau:

> I confess that we are made for each other and that this mutual dependence results in all the advantage of society. But the first intention of this union, or Social Contract, was to oblige all the contracting parties to lend each other mutual aid, and not to allow some to usurp everything; to authorise them even in their usurpations & to allow others to lack everything.[37]

This makes it clear that Antoine-Joseph Pernety contrasts Cornelius de Pauw's damning assessment of the indigenous population not only with the strong natural man and natural philosopher on the individual level but also with the man of the community, the man of the *contrat social*, on the collective level in an almost ideal-typical—and any case idealizing—manner. One may very well recognize in this a weakness of Pernety's argumentation strategy since he all too often simply opposes de Pauw's negatively colored images with positively colored counter-images.

This inverse representation, as it were, makes his argumentation dependent on that of his opponent, since Pernety repeatedly restricts himself to transforming the negative signs into positive ones at all levels. The independence of his observations suffers from this, as does the originality and flexibility of his *dissertation*, which at times degenerates into a pure counter-discourse. De Pauw made use of this strategy.

The line of argument pursued by Pernety, to assert the exact opposite of what de Pauw presented in his *Recherches philosophiques*, also characterizes the 'Seconde Partie' of Pernety's academy speech, which already on the first lines turns away from what one would have to believe 'according to Mr. de P.'.[38] For this part of the earth was the best in the world, the land was extremely fertile, the trees overloaded with fruit, as he had seen with his own eyes in the garden of the governor of Montevideo[39]—and in general, this wonderful part of the earth could best be compared with the Earthly Paradise[40] and with the gardens of the Golden Age: A world of wonders and abundance, as a Vergil had once sung about.[41]

If one were to follow de Pauw, one would have to understand the American hemisphere as a "cursed land";[42] in reality, however, Pernety had been able to convince himself that in America "the principle of life", consequently the life principle and

---

[36] Ibid.

[37] Ibid., p. 115 f.

[38] Ibid., p. 32.

[39] Ibid.

[40] Ibid., p. 35.

[41] Ibid., p. 36.

[42] Ibid., p. 43.

probably also the life force, were much more pronounced than in Europe.[43] None of what de Pauw asserts in his *Recherches philosophiques* persists in Pernety's counter-discourse. But was this an intelligent discourse strategy?

It was undoubtedly as accurate as it was wise, against de Pauw's one-sided accusations that the American continent had overrun the rest of the world with epidemics and diseases—first and foremost syphilis—to draw attention, with good reason, to reciprocal globalization of diseases, such as had indeed accompanied the first phase of accelerated globalization. The inhabitants of America were by no means plagued by an infinite number of terrible diseases, but were rather to be regarded as exceedingly healthy and robust:

> Either by exercise or by the constitution of their temperament, they are very healthy, free from paralysis, dropsy, gout, hethysia, asthma, gravel, and stone; diseases which Nature, which has given so much to our continent, has still been willing to favour. She had, however, left pleurisy in Canada; and we brought them smallpox. The Americans have given us theirs by right of exchange and trade.[44]

Against the thesis of the degeneration of the human race in the tropics of the New World, which was certainly not only advocated by de Pauw, Pernety sets a hemispheric construction that puts American (i.e. indigenous) man from the far north to the deep south in the most brilliant light with reference to the most diverse travelers:

> If we go up from the north to the southern extremity of the new continent, all the peoples we meet on our way offer well-constituted men. Such are, if we are to believe Vincent le Blanc & other Travelers, the Mexicans, the Brazilians, the Peruvians, those of Paraguay, Chile & finally the Patagonians.[45]

Not only individual peoples, but all Americans were distinguished by a far greater and more honest hospitality than Europeans, and even displayed virtues that one could not fail to admire:

> They are naturally frank, truthful, and have given in all times marks of candor, courtesy, friendship, generosity, and gratitude. Those who have practiced them for a long time do them more justice than Mr. de P. If we find among them today lies, perfidy, treachery, libertinism, and several other vices, we must blame the pernicious examples of the Europeans, and the ill-treatment which the latter has exercised against them. On every page of the reports, we see how much those of the old Continent have displayed in the new, the art which they know so well, of vile deception.[46]

In view of such a picture of morals, which sharply contrasts the robust and good man of nature with the Europeans, who are not only depraved but even more depraved of other peoples, it is not surprising that Pernety also declares the

[43] Ibid., p. 42.
[44] Ibid., p. 48.
[45] Ibid., p. 49 f.
[46] Ibid., pp. 81–83.

inhabitants of America to be the happier people.[47] The little detailed argumentation strategy, which aims solely at refuting de Pauw's assertions by counter-assertions, ultimately affects large parts of the *dissertation*, which in these long passages proves to be neither empirically nor epistemologically superior to de Pauw's explanations. This had consequences for the debate, which was initially based in and emanated from Berlin.

With the question about the Patagonians repeatedly raised by Pernety the member of the Berliner Akademie (Berlin Academy) takes up a problem which also Jean-Jacques Rousseau had mentioned not by chance ("without forgetting the real or fake Patagonians")[48] in order to finally be able to bring it to a scientific investigation and final evaluation two and a half centuries after Pigafetta's first reports of giants in Patagonia. Not for nothing had Pernety added a *Relation sur les Patagons* to his travel report, which he already included in the title of his *Journal historique* with public appeal—and had not Pierre Moreau de Maupertuis himself, the president of the Berlin Academy, let it be known about the giant Patagonians that one could no longer reasonably doubt their existence?[49]

It should hardly come as a surprise that he not only accused de Pauw's assertion that the talk of the giant-sized people of Patagonia were mythical creatures fabricated by European travelers of partial manipulation of sources[50] but countered it in a very fundamental manner:

> I cannot conceive how Mr. de P. undertook to destroy the existence of the Patagonian Geánts. Reasoning according to his philosophical method, nothing was more capable than this existence, of proving to his eyes, the degradation & degeneration of the human race in America.[51]

But it is precisely at this point that it becomes clear how much Pernety overshot his target with his not infrequently very mechanical effort to refute de Pauw on all points and damaged the legitimacy and authority of his discourse fundamentally. This may have contributed decisively to the indisputable fact that Cornelius de Pauw's *méthode philosophique*, incriminated by Pernety, was ultimately able to prevail in the Berlin debate, i.e. in that *European* phase of the dispute about the New World that initially emanated from Berlin.

Already the first rounds of this dispute went without question to Cornelius de Pauw. For the latter had only waited for the chance of a counterattack, which essentially sought to reverse his devaluation of the Americans into an idealizing counter-image of everything American, and generously printed Perney's attack in his

---

[47] Ibid., p. 87.

[48] Rousseau, Jean-Jacques: *Discours sur l'origine et les fondements de l'inégalité parmi les hommes*, p. 213.

[49] See Moureau, François: L'Amérique n'a aucun avenir: les idées 'philosophiques' de Cornelius De Pauw. In: *Travaux de Littérature* (Genève) 24 (2011), p. 66.

[50] Pernety, Antoine-Joseph: *Dissertation sur l'Amérique et les Naturels de cette partie du Monde*, p. 68.

[51] Ibid., p. 51.

three-volume new edition of his *Recherches* already mentioned and still published in 1770 again in Berlin. To it, he added a vehement and yet well-calculated defense of his own, in many cases further sharpened theses. Pernety's over long stretches inverse representation of the theses of his opponent could be easily discursively appropriated by the latter.

Certainly, one may recognize in Pernety "elements of a modern ethnological attitude",[52] among which one can certainly count "the critique of Eurocentrism; the importance of gathering data in the field and verifying sources; the recognition of the cultural diversity of the other".[53] But his way of argumentation remained—as we have already seen—too much bound to the theses put forward by de Pauw, which Frederick II's librarian often only gave a different sign, without giving them a new argumentative (and scientifically justifiable) foundation.

Thus, de Pauw became the man of the hour—and far beyond. Regardless of all further publications by Pernety, which admittedly hardly contributed anything essential to the debate,[54] his work was to remain for many years at the center of the heated and bitterly fought worldwide debates about America and the Americans, before it then fell into what was oblivion until the recent past. Not for nothing, however, did the französische Nationalversammlung (French National Assembly), in a decree of August 26, 1792, declare Cornelius de Pauw a *citoyen français*, along with other personalities deserving of the freedom of peoples, including George Washington.[55]

And yet, perhaps the decisive front of resistance to de Pauw's ideas developed from Berlin itself. For it would be easy to prove that the positions put forward in the Berlin debate were already dealt a scientific death blow by the writings published by the *philosophe voyageur* Alexander von Humboldt a few years after his return from his journey to the American tropics (1799–1804). Thus, in his introduction to his *Vues des Cordillères et Monumens des Peuples Indigènes de l'Amérique*, dated Paris, April 1813, the well-traveled scholar noted with great clarity:

> The ardour with which one had been engaged in researching America diminished from the beginning of the seventeenth century; the Spanish colonies, which enclose the only regions formerly inhabited by civilised peoples, remained closed to foreign nations; and recently, when the abbot Clavigero published in Italy his Ancient History of Mexico, one regarded as very doubtful the facts attested by a host of eyewitnesses, often enemies of one another. Famous writers, more struck by the contrasts than by the harmony of nature, took pleasure in depicting the whole of America as a marshy country, contrary to the multiplication of animals, and newly inhabited by hordes as uncivilised as the inhabitants of the South Sea. In historical research on the Americans, absolute scepticism had been substituted for healthy criticism. The declamatory descriptions of Solis and some other writers who had

[52] Mannucci, Erica Joy: The savage and the civilised: observations on a dispute between an enlightened writer and an illuminist. In: *Studies on Voltaire and the Eighteenth Century* (Oxford) 303 (1992), p. 384.

[53] Ibid.

[54] See among others Pernety, Antoine-Joseph: *Examen des Recherches philosophiques sur les Américains et de la Défense de cet ouvrage*. 2 vols. Berlin: G.J. Decker, Imprimeur du Roi 1771.

[55] See Moureau, François: *L'Amérique n'a aucun avenir*, p. 68.

> not left Europe were confused with the simple and true accounts of the first travellers; it seemed the duty of a philosopher to deny everything that had been observed by missionaries.
>
> Since the end of the last century, there has been a welcome revolution in the way we look at the civilisation of peoples and the causes that prevent or promote progress. We have come to know nations whose manners, institutions and arts differ almost as much from those of the Greeks and Romans as the primitive forms of destroyed animals differ from those of the species which are the object of descriptive natural history. The Calcutta Society has shed a bright light on the history of the peoples of Asia. The monuments of Egypt, described today with admirable accuracy, have been compared with the monuments of the most distant countries, and my research on the indigenous peoples of America appears at a time when everything that departs from the style of which the Greeks have left us, inimitable models, is not considered unworthy of attention.[56]

This passage marks an admittedly temporary conclusion to that Berlin debate in which Antoine-Joseph Pernety, 'Le Philosophe la Douceur' alias Zaccaria de Pazzi de Bonneville,[57] Giovanni Rinaldo Carli, Fray Servando Teresa de Mier y Guerra, Delisle de Sales, Francisco Javier Clavijero, George Washington, Drouin de Bercy and many others had intervened. With unmistakable allusions to the works of de Pauw and Raynal, Alexander von Humboldt not only signalled the emergence of a new discourse on the New World, in the history of which Clavijero's name could not be omitted but also drew attention to the fact that new, empirical bases for the circulation of knowledge about the entire non-European world had emerged with a view to both 'the West Indies' and 'the East Indies'. The consciousness of presenting his work in a veritable epochal threshold and of contributing something substantial to it could hardly have been more pronounced in Alexander von Humboldt. Once again, then, in this debate about the New World, the essential impulses emanated from Berlin and Prussia.

The aspects listed in dense succession in the passage cited were, moreover, placed by Humboldt in the cheerful semantics of a 'happy revolution' (which differs profoundly from the "étonnante révolution" ("an astonishing revolution") of which de Pauw spoke at the beginning of his work).[58] Step by step, new documents in the archives and libraries of both the New and the Old World, but also the field research of many American and European scholars on the spot, invalidated the now untenable theses of the Dutch philosopher whose name hardly needed to be mentioned in Humboldt's writings.

The Berlin debate had led to a worldwide dispute, the outcome of which contributed significantly to the scientific revolution observed and documented by Humboldt. Without the experience of the second phase of accelerated globalization and its consequences, this scientific revolution would not have taken place in such a resounding manner.

---

56 Humboldt, Alexander von: *Vues des Cordillères et Monumens des Peuples Indigènes de l'Amérique*. Nanterre: Editions Erasme 1989, p. 96, p. II f.

57 See Zantop, Susanne: *Colonial Fantasies*, p. 47.

58 Pauw, Cornelius de: *Recherches philosophiques sur les Américains, ou Mémoires intéressants pour servir à l'Histoire de l'Espèce humaine*, vol. I, p. a2v.

All this should not, of course, lead us to underestimate the impact of Cornelius de Pauw's theses in particular. For de Pauw's main work undoubtedly had a diffuse, but long-lasting and not infrequently subcutaneous effect, which can by no means be determined by the *explicit* reference to the name of the Amsterdam-born. It would therefore by no means be an exaggeration to speak, in further development of Antonello Gerbi's[59] considerations, of a diffuse long-term effect, which could be related in particular to philosophy, but also the area of tension between philology and racism.

For de Pauw's thinking on inferiority remained alive in Europe for a long time, not only in Hegel's famous *Vorlesungen über die Philosophie der Weltgeschichte*, but also with regard to the non-European world. Today, the insight into the historicity of the Berlin debate is linked to the necessity of reviewing its historicity polylogically and relationally, critically and self-critically, in the light of the demands of *our* phase of accelerated globalization, in order to make use of the constructive, creative—and not the destructive—power of the sciences.

In doing so, it is important to understand the Enlightenment as a global epochal phenomenon in a very fundamental way, whereby one should resist all temptations to schematically oppose representatives of the Enlightenment of European or American origin and, disregarding the actual complexity of the discussions, discourses and debates, to evaluate them positively or negatively depending on the perspective adopted. Even with this challenge in mind, the Berlin debate is far from over. However, in Berlin, where this debate started and from where, in the eighteenth century, only rarely did a question arise that was discussed worldwide, this Berlin debate about the New World has been largely forgotten. It is time, however, to recall this phenomenon, which has become historical, in order to examine and understand the contemporary demons of racism and colonialism in their development in a form that is updated today. For the perpetuators of de Pauw's theses have not disappeared.

[59] See Gerbi, Antonello: *La Disputa del Nuovo Mondo*, p. 118 f.

Alexander von Humboldt or from Revolution to Revolution

'Paris croisé'—Revolution and Circulation

In the third volume of his work on the history of the discovery of the New World, published in French in Paris between April 1834 and August 1838, Alexander von Humboldt had a footnote printed in which he dealt with the sources for Christopher Columbus' daring calculations of the end of the world. In doing so, he traced over several pages the readings of the European explorer and the complex textual filiations leading via the *Alphonsine Tables* to the writings of Cardinal Pierre d'Ailly, according to whose calculations one of Saturn's great periods was to have ended in 1789.[1] According to Humboldt, the cardinal was not able to say in 1414 whether the world would survive the resulting changes, which would particularly affect the laws.[2] This arrival of the Antichrist, expected based on astronomical signs for the year 1789, was commented on by Humboldt—as so often deliberately leaving the point in time and space with which he was concerned in this work—as follows:

> One wonders if this coincidence of dates, this prediction of a revolution that occupies such a great place in the history of the human race, had not already been signaled by those who are nowadays fond of everything that is mystical and tangible.[3]

This prediction of a revolution that would occur centuries later and the commentary on it by the author of a five-volume work, at the centre of which is the study of the history of the discovery of America and thus of the first phase of accelerated globalization, carried out with the utmost care and historical as well as philological precision, is revealing in several respects. For, on the one hand, one might well venture the assumption that Alexander von Humboldt deliberately smuggled the reference

[1] Humboldt, Alexander von: *Examen critique de l'histoire de la géographie du Nouveau Continent et des progrès de l'astronomie nautique aux quinzième et seizième siècles*. 5 vols. Paris: Librairie de Gide 1836–1839, here vol. 3, p. 256 f.

[2] Ibid., p. 257.

[3] Ibid.

O. Ette, *Mobile Prussia*, https://doi.org/10.1007/978-3-476-05839-3_4

to the French Revolution into his *Kritische Untersuchung* of the Expansion of Europe, since he was thereby giving his reading public to understand that one should certainly grant the events of 1789 a similarly prominent place in the history of the human race as those developments that led to the discovery and invention of a New World. And on the other hand, the scholar, who was living at the Prussian court at the time—where people did not like to hear about revolutions—was not only referring to an event that was to be erased from the collective memory of the numerous followers of a 'mystical and mysterious darkness'[4] in Europe at the time but which had been of the greatest importance for his intellectual development: The end of the *Ancien Régime* and the paradigmatic revolution of European modernity.

For at the end of his journey to the Lower Rhine, Holland, England and France—and thus to three European protagonists of accelerated globalization—undertaken from March 25 to July 11, 1790, together with none other than Georg Forster, the younger of the two Humboldt brothers had become acquainted with a revolutionary Paris that captivated him from the outset. The French capital, which he saw for the first time and from the perspective of the revolutionary-minded Forster, who had accompanied James Cook on his second circumnavigation of the world and written a celebrated travelogue about it that strongly influenced Humboldt, fascinated him just as much as the unmistakable movement and excitement of the masses that filled the streets of Paris. The young man, who later described the enthusiasm with which he had lent a hand, carting sand for the erection of the "still unfinished Temple of Liberty […] had carted sand"[5] and absorbed images that were to accompany him throughout his life and stand "before his soul".[6] From the very beginning, Paris offered Alexander the spectacle of a world quite different from Berlin.

For Alexander von Humboldt, the capital of the nineteenth century, which was already taking shape, was not only the city of freedom he had fought for in a revolutionary way and the outstanding centre of the sciences and arts but also the place from which he hoped to realize his dream of travelling the world.

When, from March 1799, Alexander von Humboldt sought permission from the Spanish King Carlos IV and his then still powerful Minister of State Mariano Luis de Urquijo at the courts of Madrid and Aranjuez to visit the Spanish colonies around the world, he wrote a brief account of his own life. In these pages, which were both concise and silent, he emphasized how much he was "inspired by the ardent desire"[7] to "see another part of the world, and that in relation to general natural history, not only to study the species and their characteristics (a study that has been too exclusively devoted to this day), but the influence of the atmosphere and chemical composition on living beings, the structure of the globe, the correspondence of the strata in the most distant countries, and finally the great harmonies in nature."[8]

---

[4] Ibid.

[5] See Beck, Hanno: *Alexander von Humboldt.* 2 vols. Vol. 1. Wiesbaden: Franz Steiner 1959, p. 27.

[6] Ibid.

[7] Humboldt, Alexander von: Mein Lebenslauf 1769–1799. In: Alexander von Humboldt: *Aus meinem Leben. Autobiographische Bekenntnisse.* Compiled and explained by Kurt-R. Biermann. Munich: C. H. Beck 1987, p. 27.

[8] Ibid., p. 27 f.

He had, therefore, asked the Prussian king to allow him to leave his service "for a few years" and to sacrifice part of his "small fortune to the progress of science".[9] Humboldt was determined to use the not entirely inconsiderable fortune inherited after his mother's death for a non-European voyage of exploration, which he planned to embark on together with the Frenchman Aimé Bonpland.

In this revealing sketch of his wide-ranging and truly global (and globalizing) research program for a voyage that was to lead him to the 'equinoctial regions of the New Continent' Humboldt, who as Prussian and Protestant (and in the company of a Frenchman) would have seemed doubly dangerous at other times at the Spanish royal court, Humboldt had 'forgotten' his lasting interest in the cultures of the 'indigenous peoples'—as he was soon to call them in the title of his *Vues des Cordillères et Monumens des Peuples Indigènes de l'Amérique*—as well as in the social, political and economic developments in the Spanish colonies.[10] Had he mentioned his interest in colonial conditions, he would never have been granted access to the Spanish overseas colonies—despite Urquijo's liberal intermediate phase, which was so favorable to Humboldt.

However, the then still barely thirty-year-old (who had long since made a name for himself internationally through lectures in Paris) did not forget that originally it was not Madrid but Paris that was to be the starting point of his planned world tour:

> The French Directory had decided to have a circumnavigation of the world made with three ships under the command of Captain Baudin; I was invited by the Minister of the Navy to combine my work with that of the scholars who were to take part in the expedition. I was already preparing to set out for Le Havre when lack of money caused the project to fail.[11]

Just as he evoked the image of a Paris of the *Directoire* in Madrid—which was not exactly popular at the Spanish royal court—Alexander von Humboldt was still to wear the clothes of the *Directoire* many years later as a chamberlain at the Prussian royal court committed to democratic principles and the ideals of the French Revolution. Still a famous figure at the Prussian royal court far beyond the borders of Prussia and Europe, Humboldt was not regarded by many as the anti-Christ—even if he was publicly insulted as a 'soul murderer' in January 1857 because of his 'godless' *cosmos*[12]—but as a disguised anti-monarchist. It was not without reason that he followed the coffins of the fallen revolutionaries in Berlin in the funeral procession on March 22, 1848. His whole habitus was a symbolic sign of his attachment not only to the French

[9] Ibid., p. 28.

[10] On the Spanish dimension of Alexander von Humboldt's journey, see also the dissertation by Rebok, Sandra: *Alexander von Humboldt und Spanien im 19. Jahrhundert: Analyse eines wechselseitigen Wahrnehmungsprozesses.* Frankfurt a. M.: Vervuert 2006; as well as the volume by Puig-Samper Mulero, Miguel Angel/Rebok, Sandra (eds.): *Sentir y medir. Alexander von Humboldt en España.* Prólogo Ottmar Ette. Aranjuez: Ediciones Doce Calles 2007.

[11] Humboldt, Alexander von: My curriculum vitae. In: Alexander von Humboldt: *Aus meinem Leben*, p. 29 f.

[12] See Biermann, Kurt-R./Schwarz, Ingo: Why did Alexander von Humboldt refer to himself as 'the old man from the mountain' (Vecchio della Montagna)? In: *Alexander von Humboldt-Stiftung. Mitteilungen, Alexander von Humboldt-Magazin* (Bonn) 60 (1992), pp. 71–73.

capital, where he lived for several decades in the course of his long life, and to the French language, in which he wrote most of his scientific and literary works, but also to his loyalty to the republican ideals of his first visit to Paris and the power of fascination that the *ville lumière* had always exercised on the world-conscious Prussian. It is still fascinating today to see how for many decades Humboldt lived and thought in a political and social context in Prussia that largely contradicted his ideals. And how he managed to remain in dialogue and mutual exchange with everyone, while at the same time not deviating from his political and social ideas.

The Paris that Alexander von Humboldt adored was not only the Paris of the sciences and his scientific friends, not only the Paris of the salons and that French rhythm of life to which Humboldt adapted so excellently after his return from the American expedition (1799–1804), not only the Paris of the French language and great publishing houses, to which he felt he belonged as much as a German as a French writer, but above all that *Paris croisé* that had become the site of a tremendous circulation of the most diverse forms of knowledge. For Alexander von Humboldt, even during those times when the French capital seemed to be dominated by reactionary political forces, Paris was the hub of a *circulation des savoirs* that was global in scope and represented the accelerated formation of what the author of *Kosmos* called world consciousness.[13]

Alexander von Humboldt's Paris is the Paris of ideas—and perhaps for that very reason always a little idealized. Above all, however, the French capital had long since become ubiquitous for him in a certain sense—Humboldt carried it around with him in his habitus and his rhythm of life, even in his clothing. From the very beginning, Paris was far more to him than a geographically localizable place: It stood for the fascination of revolution.

Ici est un autre

When Alexander von Humboldt spent long periods in Potsdam during the last third of his life—i.e. after the end of his Russian-Siberian expedition in 1829—he always liked to furnish the Prussian residential city with elements of his American expedition in his characteristic ironic and self-deprecating manner. For example, Potsdam's Brauhausberg, which was Humboldt's home mountain for many years and remained so until his old age, and which he still liked to climb with his friend Berghaus, was stylized by him as his "Potsdam Chimborazo,"[14] while the Havel River was renamed the "Magellanic Strait"[15] at the point where it spanned the so-called Long Bridge in the immediate vicinity of Potsdam's City Palace—where Humboldt could have easily heated rooms on the mezzanine floor. Humboldt loved to project American travel images onto Potsdam and thus Americanize the Prussian residential city.

[13] See Ette, Ottmar: *Weltbewußtsein. Alexander von Humboldt und das unvollendete Projekt einer anderen Moderne.* Weilerswist: Velbrück Wissenschaft 2002.

[14] See ibid., p. 58.

[15] Ibid.

Were these the moods of an old man or—as Humboldt once described himself in his later years, in a letter of December 3, 1856, to his friend Varnhagen von Ense—of a "super old man with a mossy head"?[16] When, in the spring of 1829, before his Russian-Siberian expedition had really begun, he and his team of explorers were stuck on the Curonian Spit for a few days because of the ice and the resulting impossibility of crossing to Memel, he began to mock his predicament and to fantasize about the other 'sandy desert' that he had just left on his journey in the landscape of sand and scrub that lay before him:

> If Schinkel had a few bricks glued together there, if a Monday Club, a circle of art-loving Jew-demoiselles and an academy were set up on those scrub-covered sandy steppes, nothing would be lacking to form a new Berlin; indeed, I would prefer the new creation, for I have seen the sun dip gloriously into the sea on the spit.[17]

At the beginning of his Russian-Siberian expedition between April 12 and December 28, 1829 within the horizon of the Russian-Siberian expedition to the empire of the tsar who had been nicknamed the "gendarme of Europe"[18] on account of his uncompromisingly authoritarian methods of rule and oppression and who had been placed by Humboldt in his dedication insidiously in the context of what had been called "the free development of the intellectual faculties",[19] Alexander's letter to his brother in Berlin gave rise to a peculiarly iridescent picture of the Prussian capital whose "unnature"[20] after the Oranienburg Gate now already extended for 200 miles to the northwest. Berlin did not seem to want to end in the East.

The Russian-Siberian expedition was accompanied by serious restrictions: In order to obtain permission and funding for this great Asian voyage, which he had dreamed of for decades, Humboldt had had to agree to the condition that he report only on 'dead nature' and not on the ailing and repressive social conditions of the giant empire.[21] How could he not have thought of his largely independently conducted American journey during this Asian one?

---

[16] *Briefe von Alexander von Humboldt an Varnhagen von Ense aus den Jahren 1827 bis 1858*, together with excerpts from Varnhagen's diaries and letters from Varnhagen and others to Humboldt. Leipzig: F. A. Brockhaus 1860, p. 341.

[17] Humboldt, Alexander von: Letter of April 29, 1829 to Wilhelm von Humboldt. In: *Briefe Alexander's von Humboldt an seinen Bruder Wilhelm*. Edited by the von Humboldt family in Ottmachau. Stuttgart: Verlag der J. G. Cotta'schen Buchhandlung 1880, p. 170. See in detail Ette, Ottmar: Amerika in Asien. Alexander von Humboldt's *Asie centrale* and the Russian-Siberian expedition in a transareal context. In: *HiN—Alexander von Humboldt on the Net. International Journal of Humboldt Studies*. VIII, 14 (2007), pp. 17–40, 10.18443/89.

[18] See on this as well as on Humboldt's criticism of the tsarist regime, among others, Suckow, Christian: Alexander von Humboldt und Russland. In: Ette, Ottmar/Hermanns, Ute/Scherer, Bernd M./Suckow, Christian (eds.): *Alexander von Humboldt—Aufbruch in die Moderne*. Berlin: Akademie-Verlag 2001, p. 249.

[19] Humboldt, Alexander von: *Asie Centrale. Recherches sur les chaînes de montagnes et la climatologie comparée*. 3 vols. Paris: Gide 1843, here vol. I, p. viii.

[20] Humboldt, Alexander von: Letter of April 29, 1829. In: *Briefe Alexander's von Humboldt an seinen Bruder Wilhelm*, p. 170.

[21] See Ette, Ottmar: Amerika in Asien, chap. "Rahel Levin Varnhagen oder das Mobile Preußens".

Just as Berlin suddenly appeared at the Curonian Spit, so in his work *Asie Centrale*, published in Paris and French in 1843, the image of Chimborazo, as well as other Andean peaks, was initially inserted in the midst of the Siberian steppes in a completely surprising way. And just as Humboldt's clothing at the Prussian royal court in Berlin or Potsdam unexpectedly embodied the Parisian period, so in the writings of this great explorer and writer other places appeared among the places, other landscapes among the landscapes. Often a peculiar oscillation made itself felt in his letters and writings, which always connected a certain place with other places, and even unawares faded distant cities into cities present to him-at first without any apparent reason. One could therefore speak, in modification of a famous autobiographical formula of Arthur Rimbaud (*Je est un autre*) of the simultaneous otherness (and Being-Somewhere-Else) of one's own: *Ici est un autre.*

Consequently, a certain place does not appear in Humboldt's work as a static space whose boundaries and territoriality can be clearly outlined and fixed, but rather as a space of movement that is characterized by often perplexing connections, but always by its integration into mobile structures and structuring. With a view to such relations, one gains some depth of focus, not only biographical, if one takes into account the autobiographical text that Humboldt had begun to write down in Cuba and which he dated Santa Fe de Bogotá, August 4, 1801. At that time, Humboldt recalled the opportunity he had been given in 1790 to accompany Georg Forster on the aforementioned journey:

> I was then ill, March 1790, at Göttingen, and engaged in the publication of my first literary product, the Basalts on the Rhine. Nevertheless, with what joy I took part in this journey. Although it distanced me more from Forster than it brought me closer to him, like any close living together among people, and especially with Forster's petty and vain character, living together with the circumnavigator had a great influence on my inclination for the tropical world. How much this longing awoke fully at the sight of the all-spreading, moving, country-connecting ocean, which I first saw at Ostend, how much at the little crossing from Hellevoetsluis to Dover.[22]

It is characteristic of Humboldt's thinking that in this passage—as in later writings—the sea is not understood as separating different landmasses from one another, but rather as connecting them with one another and as the *mobile* element *par excellence*. In the autobiographical review, Humboldt observes how his "warm imagination fills with longed-for figures of distant things" after his mind—according to the freedom-loving traveller—had been "maltreated for eighteen years in his father's house and confined to a meagre sandy nature".[23] In his room in England, he is fascinated by the image of an "East Indian ship" that "sank in a storm"[24]—and even later, the idea of his shipwreck would always remain alive, despite all the desire for long-distance travel.

[22] Humboldt, Alexander von: Ich über mich selbst (my path to becoming a natural scientist and explorer 1769–1790). In: Alexander von Humboldt: *Aus meinem Leben*, p. 36.

[23] Ibid., p. 38.

[24] Ibid.

The young Humboldt lives as if in a dream as if somewhere else. He believes himself to be "anchored in Cape Town", but "with rising sun" the "sweet dream was wiped away".[25] Everything arouses the "desire for the tropics"[26] in him: "A desire like this, which accompanied me eternally, the aspiration to lands where we are separated from our own by boundless spaces."[27] And Humboldt concluded the reflections of his one's-own-life-writing (EigenLebenSchreibens) with the following sentences:

> I often cried without knowing why, and poor Forster agonized to find out what lay so darkly in my soul. With this mood, I returned to Mainz by way of Paris. I had made distant plans.[28]

In these passages too, which reflect the experiences of a young traveller, another place always appears beneath a place, another country, another landscape always becomes visible in the presence of a landscape, a country: *Ici est un autre*. The fascination of distance becomes comprehensible in the wanderlust of a young man who, in the company of a world traveller, finds himself on a journey through Europe, but above all longs for another world of distant lands and seas.

If we jump from the level of the narrated I to that of the narrating I, it becomes clear that the writing takes place from the very voyage around the world of which the narrated I once dreamed. Behind Forster's and Humboldt's joint voyage to England and France, therefore, not only *en amont* Forster's circumnavigation of the globe with Cook but also *en aval* Humboldt's American voyage of exploration with Bonpland becomes present. The voyages overlap: Beneath one voyage, a completely different one becomes visible, which in turn refers to future voyages that have themselves long since become realized voyages. Humboldt's autobiographical long-distance writing from movement unfolds a complex interweaving of (travel) movements: *Un voyage peut en cacher un autre.*

In Alexander von Humboldt's *EigenLebenSchreiben* it becomes apparent that the author's experience of space and his conception of space is based on a knowledge of life and an experience of life that is never derived from a single place. A place is always inscribed with the ‚Fort' (away), the supposedly separating sea always with more connecting movement.

By the time of his American research trip, Alexander von Humboldt had long since developed a complex theory of space and movement. In the preceding considerations, I was therefore not concerned with anchoring his spatial conceptions solely in a life experience and resolving the Humboldtian epistemology into a series of revealing, but not conceptually well-thought-out biographies. Rather, it is crucial to understand Humboldt's epistemology as a structure and structuring—incidentally, still highly demanding and forward-looking even for our time—that must not disappear behind any biography, no matter how moving.

---

[25] Ibid., p. 39.

[26] Ibid.

[27] Ibid.

[28] Ibid., p. 40.

But without the vital contexts, without the biographies, Humboldt's episteme cannot be thought of in its genesis: Lifestyle and style of thought are too inextricably intertwined in the author of *Kosmos*. Travel and movement have been part of Humboldt's transdisciplinary and intercultural science from the very beginning.

The Port of Havana and the Island as a Space of Movement

We had seen that Alexander von Humboldt understood the sea very early on as the mobile, moving and connecting element in itself. In this respect, the author of the *Geography of Plants* repeatedly took special care in his travel work to create scenes of departure or arrival by ship. This is because they form highlighted travel-literary places that have a particularly high semantic density and significance on a literary level. Port cities have specific relevance in this context, as they open up space for which they stand as gateways.

The Cuban writer and *poeta doctus* Alejo Carpentier, therefore, made an excellent choice when he chose as *incipit* for his literary declaration of love for Havana—which, contrary to the legend he often spread, was not his birthplace—the description of the entry into the port of the Cuban capital that Alexander von Humboldt had printed in the 28th chapter of his *Relation historique*.[29]

This chapter of his report on the American voyage of discovery, which remained fragmentary and was published in French in Paris between November 1814 and April 1831, appeared in the third volume of *Relation* and was published parallel to the travel report as a separate print under the title *Essai politique sur l'île de Cuba* in 1826. The entry into the port of Havana, which Humboldt enjoyed for the first time on 19 December 1800, is spectacularly staged:

> The aspect of Havana, at the entrance to the port, is one of the most rians and picturesque to be enjoyed on the coast of equinoctial America, north of the equator. This site, celebrated by travellers of all nations, has not the luxury of vegetation which adorns the banks of the river of Guayaquil, nor the savage majesty of the rocky coasts of Rio Janeiro, two ports of the southern hemisphere: But the grace which, in our climates, embellishes the scenes of cultivated nature, is mingled here with the majesty of vegetable forms, with the organic vigour which characterises the torrid zone. In a mixture of such sweet impressions, the European forgets the danger which threatens him in the midst of the populous cities of the West Indies; he seeks to grasp the various elements of a vast landscape, to contemplate these castles which crown the rocks to the east of the port, this interior basin, surrounded by villages and farms, these palm trees which rise to a prodigious height, this city half-hidden by a forest of masts and the sails of the ships.[30]

The literary composition of this impressive passage is based, as it were, on a narrative *travelling*, since the port and the city of Havana unfold from the perspective of

[29] See Carpentier, Alejo: *La ciudad de las columnas*. La Habana: Editorial Letras Cubanas 1982.

[30] Humboldt, Alexander von: *Relation historique du Voyage aux Régions équinoxiales du Nouveau Continent [...]*. Reprint of the 1814–1825 in Paris of the published complete original, procured, introduced and augmented with an index by Hanno Beck. Vol. III. Stuttgart: Brockhaus 1970, p. 348.

a traveller observing from onboard, who portrays the place of movement of the main Cuban port from his movement. Alexander von Humboldt often applied this principle of writing and portraying from movement: Even in his *Summa*, the *Kosmos*, the reading public is led, as it were, along cosmic paths through space to the globe and onto the surface of our planet.

In the context of our considerations, we are not concerned here with the specifically literary forms of *Humboldtian writing* as writing (in) modernity.[31] The contrasting position of this 'picturesque' port, frequently visited and 'celebrated' by travelers from all over the world, in relation to the ports of Guayaquil (which Humboldt knew) and Rio de Janeiro (which was known to him only through engravings and descriptions in travel literature), leads in this passage to the determination of a specific intermediate position of Havana. For in this port the climatic conditions of the temperate zone and the 'cultivated nature' of the North intersect on the one hand with the 'majesty' of tropical plants as they characterize the hot zone.

Thus, the port of Havana already appears at this level as a place of exchange, a circulation that obviously connects the North with the South and at the same time the East with the West, Europe with America. In this way, the necessity of understanding a space from the point of view of movement is already indicated, namely from its port, its main port, which makes communications and the resulting circulation of knowledge possible in the first place.

To the mixture of climates and of plants corresponds a 'mingling' of such sweet and pleasant impressions that they hide from the European traveller, sensually bewildered and beguiled by them, the danger which is no less from the circulation in these densely populated cities of the Antilles. The contrasts of nature and culture, of city and country, of mobile water and solid rock, are carefully interwoven and blended, so that a moving and at the same time stirring scenery unfolds, which will be further elaborated in the sequel. Nature and culture are thought together: We are undoubtedly dealing here with a literary painting of nature by Alexander von Humboldt.

It would be easy to prove that the entry into the port of Havana functions like a *mise en abyme* of the entire text. Relevant to our question is, therefore, the fact that Alexander von Humboldt, through the mobile perspectivization he has chosen, stages the port of Havana as a *fractal* pattern[32] of the entire island from the very beginning as a place of movement, where highly contrasting climates, plant forms, cities and populations meet and a circulation of goods as well as of people—and also of people as goods, since the view is directed early on to the sale of the "unfortunate slaves"[33]—of cultures and languages, of traditions and ideas begins. The

[31] See Ette, Ottmar: Eine 'Gemütsverfassung moralischer Unruhe'—'Humboldtian Writing': Alexander von Humboldt und das Schreiben in der Moderne. In: Ette, Ottmar/Hermanns, Ute/Scherer, Bernd M./Suckow, Christian (eds.): *Alexander von Humboldt—Aufbruch in die Moderne*, pp. 33–55.

[32] On the epistemological significance of the fractal for literary studies, see Ette, Ottmar: *WeltFraktale. Wege durch die Literaturen der Welt.* Stuttgart: J. B. Metzler 2017.

[33] Humboldt, Alexander von: *Relation historique*, vol. III, p. 350.

epistemological status of this passage is obvious: Wherewith Cornelius de Pauw or Antoine-Joseph Pernety the opposites faced each other more or less statically, with Alexander von Humboldt they are intensely related to each other and brought into constant and mutually changing exchange with each other thanks to that "sight[s] of the all-spreading, moving, country-connecting ocean".[34] The port of Havana is a place of transit that can only be understood from its mobile relationality, not from its static territoriality.

Certainly, nothing could better convey the island of Cuba, which played a weighty role in the first as well as the second and third phases of accelerated globalization, than a representation that makes the largest of the Antillean islands appear as an intense transoceanic space of movement. The tropical profusion of palm trees on land is matched by entire forests of masts whose sails represent the abundance of human-initiated forms of circulation. It is no coincidence that the description of the harbor of Havana recalls the description of the harbor of Amsterdam in that travelogue which Georg Forster sketched of the common voyage in his *Ansichten vom Niederrhein*. But more about this in the last chapter of the present volume.

The epistemological quality of this literarily successful arrival is thus evident: Humboldt's understanding of Cuba is not based on a comparison between two fundamentally separate and distinct sides of the Atlantic but works out a mobile relationality beyond comparison, in which everything seems connected to everything else. The theorem that is perhaps central to Humboldtian thought—"Everything is interaction"[35]—refers not only to the reciprocal transformations of nature and nature, culture and culture, or nature and culture but aims at a global space of constant exchange that is opened up by the ocean connecting countries as a mobilizing force, as a basic element of dynamics.

That this accelerated movement could not only produce worldwide exchange relations but also wash away the foundations of Cuban slaveholding society itself—the experience of the Haitian Revolution still raging on the neighboring island at the time of Humboldt's two stays in Cuba had not faded away even at the time of the writing of these passages—was highly conscious to the author of the *Essai politique sur le royaume de la Nouvelle-Espagne*. The revolutions and upheavals that could be triggered by political, social or other emancipatory movements were part of Humboldt's concept of movement and knowledge of life.[36]

Against the background of the revolutionary events in Saint-Domingue, formerly the richest European colony in the Antilles, but also in view of the French Revolution and his own experience of revolutionary Paris, Humboldt knew how quickly

[34] Humboldt, Alexander von: *Ich über mich selbst*, p. 36.

[35] Humboldt, Alexander von: *Reise auf dem Río Magdalena, durch die Anden und Mexico.* Part I: Texts. Compiled from his travel diaries and explained by Margot Faak. With an introductory study by Kurt-R. Biermann. Berlin: Akademie-Verlag 1986, p. 358.

[36] On the dimensions of life knowledge in Humboldt's American Travel Diaries, see the edition of Humboldt, Alexander von: *Das Buch der Begegnungen. Menschen—Kulturen—Geschichten aus den Amerikanischen Reisetagebüchern.* Edited, translated from the French and annotated by Ottmar Ette. With original drawings by Humboldt and historical maps and timetables. Munich: Manesse 2018.

structures that had always been thought stable could change abruptly under the influence of ideas once they had been put into circulation. Thus, he wrote with the rapid demographic developments with Cuba in mind:

> If the legislation of the Antilles and the state of the people of the colour do not soon prove to be salutary changes, if we continue to discuss without taking action, political responsibility will fall to those who have the strength to work, the will to fight and the courage to end long privations. This catastrophe has taken place as a necessary consequence of the circumstances, and without the free blacks of Haiti being aware of it, without them abandoning the system of isolation that they have followed until now. Who is to predict the influence that a Confederation of African Free Antilles States, placed between Colombia, North America and Guatemala, will exert on the politics of the New World?[37]

The *Historia*, Humboldt knew, was no longer the *Magistra Vitae*.[38] Who wanted to predict the future of a revolution or of this 'African confederation' brought into play by Humboldt—which, interestingly, in Spanish translations tends to be translated as "confederación americana"? The successful revolutions in the English colonies of North America, in France and on Saint-Domingue had shown Humboldt, writing in French from Paris, how open the history(ies), how plural the futures were.

For this insight, underpinned by an epochal experience, there was perhaps no better place than Paris, the city that was Humboldt's actual (if not exclusive) centre of life during the second phase of his life, between the American expedition (1799–1804) and the Russian-Siberian expedition (1829). For it was from here that not only the global circulation of goods, people and ideas could be grasped, but also history as a complex space of movement with its wide-ranging paths of knowledge.

Space as Movement

In a phase of accelerated globalization, the phenomenon of worldwide interdependence could no longer be epistemologically addressed by comparisons between different spaces alone. Alexander von Humboldt understood that static comparisons had to be raised to the level of mobile relationalities if one wanted to understand scientifically what Georg Forster had certainly seen with regard to the development of world trade. Under the impression that "the whole history of the world" was about to dissolve into the trade of the European powers, Forster saw the duty of the researcher in taking into account "as many ideas and facts" as possible and "collecting them into a focal point".[39] But was everything really still to be concentrated in *one* focal point, from one point of view?

[37] Humboldt, Alexander von: *Relation historique*, vol. III, p. 389.

[38] See Koselleck, Reinhart: Historia Magistra Vitae. On the dissolution of the topos in the horizon of modern moving history. In: Reinhart Koselleck: *Vergangene Zukunft*. Frankfurt a. M.: Suhrkamp[2] 1984, pp. 38–66.

[39] Forster, Georg: Die Nordwestküste von Amerika, und der dortige Pelzhandel. In: Georg Forster: *Sämtliche Schriften, Tagebücher, Briefe*. Vol. V: Kleine Schriften zur Völker- und Länderkunde.

For Alexander von Humboldt's model of science, the metaphor of the burning mirror could no longer serve. Moreover, he had recognized that a fundamental change had taken place in the sciences around 1800, as he succinctly stated in his introduction to his *Vues des Cordillères et Monumens des Peuples Indigènes de l'Amérique*, dated Paris, April 1813:

> The ardour with which research on America had been carried on diminished from the beginning of the seventeenth century; the Spanish colonies, which contain the only regions formerly inhabited by civilised peoples, remained closed to foreign nations; and recently, when the Abbot Clavigero published in Italy his Ancient History of Mexico, facts attested by a host of eye-witnesses, who were often enemies of one another, were regarded as very doubtful. Famous writers, more struck by the contrasts than by the harmony of nature, took pleasure in depicting the whole of America as a marshy country, contrary to the multiplication of animals, and newly inhabited by hordes as uncivilised as the inhabitants of the South Sea. In historical research on the Americans, absolute scepticism had been substituted for healthy criticism. The declamatory descriptions of Solis and some other writers who had not left Europe were confused with the simple and true accounts of the first travellers; it seemed the duty of a philosopher to deny everything that had been observed by missionaries.
>
> Since the end of the last century, a happy revolution has taken place in the way of looking at the civilisation of peoples and the causes which stop or favour its progress. We have come to know nations whose manners, institutions and arts differ almost as much from those of the Greeks and Romans as the primitive forms of destroyed animals differ from those of the species which are the subject of descriptive natural history. The Calcutta Society has shed a bright light on the history of the peoples of Asia. The monuments of Egypt, described today with admirable accuracy, have been compared with the monuments of the most distant countries, and my research on the indigenous peoples of America appears at a time when everything that departs from the style of which the Greeks have left us, inimitable models, is not considered unworthy of attention.[40]

In this long, but at the same time very condensed passage on the history and theory of science, which is cited a second time in this volume due to its importance, Alexander von Humboldt first settles accounts—without naming names—with de Pauw, Pernety or Raynal, the *philosophes* and representatives of a meanwhile outdated epistemology, who had dealt with everything in contrasts ("contrastes"), had no empirical basis for their knowledge and had not presented any well-founded criticism beyond nonsensical generalizations. But a new time, a new epoch had dawned.

For this situation had changed fundamentally at the turn of the nineteenth century: Humboldt believed he could state a 'happy revolution'—consequently another revolution, this time, however, in the field of science—which was based, on the one hand, on empirical foundations and, on the other, had at its disposal a body of data collected worldwide. These changes indeed formed the basis for Humboldt's new discourse on the New World: A discourse based on a revolution.

Edited by Horst Fiedler, Klaus-Georg Popp, Annerose Schneider and Christian Suckow. Berlin: Akademie-Verlag 1985, p. 395.

40 Humboldt, Alexander von: *Vues des Cordillères et Monumens des Peuples Indigènes de l'Amérique.* Nanterre: Editions Erasme 1989, p. II f.

At this point, at the latest, it becomes clear that the second phase of accelerated globalization in the course of the second half of the eighteenth century had led to a fundamental change in that investigations and findings were now arriving from the most diverse parts of the world and were waiting to be critically correlated with other data sets. “Everything is interaction”—this Humboldtian theorem applied not only to nature but also and especially to the most diverse cultural phenomena, which the author of *Kosmos* took on.

This made possible a new dynamic relationality to whose accelerated interconnectedness Humboldt himself sought to contribute. To this end, he now strengthened the network of correspondents he had already built up from Paris before his American research trip, which allowed him not only to maintain but also to considerably intensify the exchange of data with specialists from various regions but also from the most diverse disciplines, even after his return to Europe. In place of dichotomous opposites, Humboldt increasingly placed concatenations and dynamic network structures, which corresponded far better to his consistent idea of a ‘harmony of nature’ in which everything had to be connected with everything else.

In the context of a flood of data that could no longer be simply bundled, but rather networked, the bringing together and even more the thinking together of information about the areas of the world most distant from each other and Europe necessitated the development of concepts that could not be limited to a single area (such as the Iberian colonies of the Americas). Thus, Humboldt’s *Vues des Cordillères et Monumens des Peuples Indigènes de l’Amérique* already demonstrate the double connection of phenomena of nature and culture on a global scale, as the title suggests, and at the same time call for forms of writing that are able to do justice to this dynamic multi-connectedness.

From today’s perspective, this concept, which Humboldt made increasingly complex in the course of a long scholarly life, can be described as a relationality that crosses individual *areas* and, consequently, as transareal. The study of the Aztec calendar stone, for example, is by no means limited to the exploration of different cultures and their conceptions of time in the high valleys of Anáhuac, the Iberian colonies or the entire American hemisphere. Rather, diverse relations are established with concepts of time and time calculations, for example, among the Tartars or Tibetans, Chinese or Japanese, among cultures of the Near or Middle East, but also the various cultures of the Occident since the Greeks and Romans.

From this point of view, Humboldt’s investigations are dedicated to a certain area, but they do not understand it ‘from itself’, but from its connections and relationships with other areas. This gives rise to a scientific approach that can certainly be attributed to the “révolution heureuse”[41] (“happy revolution”) already outlined in the sciences, but which could also be understood as an early form of the emergence of *TransArea Studies*. But what is meant by this from today’s perspective?

[41] Ibid, p. II.

Vector Dimension and TransArea Studies

If one attempts to think further about this important dimension of Humboldtian science from today's perspective, epistemological and scientific-theoretical challenges arise which will be discussed in more detail below.

In Alexander von Humboldt's concept of science, the phenomena studied are focused from the perspective of their dynamics, paths and displacements. This applies equally to his plant geography studies (which always included the 'migrations' of plants) and to his geological and mineralogical investigations (which allowed him to predict diamond and platinum discoveries in the Urals on the basis of global correlations), to his climatological research (in which he increasingly included climate changes triggered by humans) and to his investigations into the worldwide circulation of precious metals.

In the broad field of cultural phenomena that he worked on, questions of the circulation of knowledge—for example, with regard to his studies of the Aztec calendar stone and the movement of knowledge between Asia and America—were also of paramount importance to him. For Humboldt, knowledge itself is always processual and radically unfinished.

Thus, Alexander von Humboldt's conception of science is already based on the thesis that spaces (as well as knowledge formations) are fundamentally defined, configured and formed by the movements that cross them. This replaces a conception of spaces oriented solely to the topographical-geographical dimension with a dynamic, vectorial perspective that presupposes a highly dynamic basic concept with regard to the most diverse phenomena to be studied in the natural and cultural sciences. In this context, the concept of the vectorial, vectorization or vectorial dimension means that historically and culturally stored patterns of movement must be substantially included in an understanding of networked processes of generation of spaces. For it is often the stored movements that repeatedly cross certain spaces in the same figure of movement that contribute decisively to the continuity of spaces in the sense of their constant reconfiguration. Humboldt, thus, possessed a highly dynamic understanding not only of the territorial but of all processes occurring in nature and culture.

Within the context of the history of science, this implies that *area studies*, which has long since fallen into crisis, must be expanded, supplemented, and transformed by *TransArea Studies*, which first and foremost scientifically address the vectoricity of spaces on a translocal, transregional, and transnational as well as on a transareal and transcontinental level.[42] It is not the determination of space by its (supposedly natural) territoriality with its respective borders and demarcations that is to be investigated, but rather the historically accumulated and unfolded networks of forms of movement and figures of movement that are to be used as a starting point and basis

[42] See in detail Ette, Ottmar: *TransArea. Eine literarische Globalisierungsgeschichte.* Berlin; Boston: de Gruyter 2012; as well as Ottmar Ette*: Writing-Between-Worlds. TransArea Studies and the Literatures-without-a-fixed-Abode.* Translated by Vera M. Kutzinski. Berlin; Boston: de Gruyter 2016.

for understanding a certain *area* relationally-dynamically by means of the networks of relationships and linkages that cross it and thereby constitute it in a mobile way, as well as for analyzing it transareally. Of course, this also means that Humboldt did not conceive of his Prussia primarily as a territorial state, but saw it as integrated into the most diverse worldwide processes. For him, this included international economic relations, for which he repeatedly drew up global maps of movement, as well as the migrations of populations that made up an important part of Prussian society in his time.

Generally speaking, this means for us today: All this not only includes a critical questioning of the power of the borders[43] of a certain *area*, such as Latin America, or the integration of this *area* into superordinate—for example continental or hemispheric—constructions[44] but also means the attempt at a fundamental transareal reconceptualization of a space previously understood primarily in terms of its internal relationality.[45] The reciprocal linking of different *areas* with one another no longer takes place via a 'mediating' and mostly centering role of Europe; rather, it is precisely movements that have often been overlooked or degraded to 'marginal phenomena' by the previous focus on Europe that are in the foreground.

The Caribbean can be considered a particularly striking example of a complexly structured *area*, which can be delineated in an almost paradigmatic way as a highly fragmented network of island worlds and island worlds, as a multi- and translingual and at the same time profoundly transcultural unity of island subspaces and the surrounding mainland fringes in the sense of a circum-Caribbean space.[46] Each island forms its world, with its climate, its agriculture, its language and culture; and complementary to this island-world, at the same time, an island-world arises in which the multi-connectedness of all islands with one another is decisive in both archipelagic and transarchipelagic terms.

A transareal understanding of this *area* is essentially based on the assumption that the Caribbean can only be understood if we dynamically include its historically highly changeable relationships, especially with the rest of the American hemisphere, with various regions of Europe and Africa, with the Arab world, with India and with China. In this context, the 'divisions' of the Caribbean that are still predominantly anchored in disciplinary terms according to their respective affiliation with European hegemonic powers—traditional approaches that have by no means been faded out, but must be given a new perspective—stand in the way of a relational-dynamic spatial conception of the Caribbean. This, however, has to include natural conditions in its concept as well as multi-, inter- and transcultural

---

[43] See Braig, Marianne/Ette, Ottmar/Ingenschay, Dieter/Maihold, Günther (eds.): *Grenzen der Macht—Macht der Grenzen. Lateinamerika im globalen Kontext*. Frankfurt a. M.: Vervuert 2005.

[44] See Birle, Peter/Braig, Marianne/Ette, Ottmar/Ingenschay, Dieter (eds.): *Hemisphärische Konstruktionen der Amerikas*. Frankfurt a. M.: Vervuert 2006.

[45] See Ette, Ottmar/Pannewick, Friederike (eds.): *ArabAmericas. Literary Entanglements of the American Hemisphere and the Arab World*. Frankfurt a. M./Madrid: Vervuert/Iberoamericana 2006.

[46] See Ette, Ottmar (ed.): *Caribbean(s) on the Move—Archipiélagos literarios del Caribe. A TransArea Symposium*. Frankfurt a. M.; New York; Oxford: Peter Lang 2008.

contexts within the dynamic framework of a sequence of accelerated globalization processes. Here, too, the prefix 'trans' stands for the crossing of spaces, times, cultures or practices.

For the systematics of disciplines, it is important in this context that an understanding of *TransArea Studies* founded in this way no longer distinguishes between a supposedly 'general' or 'systematic' theory formation, which as a rule is very much based on the investigation of phenomena originating from certain privileged sub-regions or *areas*, and questions that are banished from the area of the 'general' theory formation of a discipline and delegated to regional scientific institutions or sub-areas.

Often the so-called 'distant competence' is institutionally misunderstood in such a way that all those competencies are gathered under it which lie far from the 'actual' centres and assumed 'cores' of the respective subjects. Thus—to take just one example—the approaches to literary theory that are regarded as essential and fundamental are almost universally based on the knowledge and analysis of very few—and mostly Western and Central European—literatures, while (with the exception of the USA) the literary-theoretical production of other *areas* is not included in 'general' theory formation.

Within a globalized theoretical and research horizon, such a separation—which mostly remains unreflected—is no longer tolerable: *TransArea Studies,* therefore, questions the prevailing mechanisms of inclusion and exclusion on the level of scientific theory and develops disciplinary and spatial concepts that are characterized far less by their territorial demarcations than by the capture of vectorial dimensions—and thus of those movements that constitute spaces as such in the first place. It would not be difficult to prove that a theoretically well-founded understanding of what constitutes European literature, beyond merely additive concepts summing up individual national literatures, is simply not possible without a transareal approach. But it is always true that the *spiritus rector* of *TransArea Studies* is the *spiritus vector.*

Paris/Berlin

But let us return to Humboldt and his Prussia. The concept of science that Alexander von Humboldt developed and shaped over seven decades was thoroughly transareal in structure, both at the level of the object and at the level of analysis in the sense outlined above. Within the emerging Humboldtian science, travel—as we have seen—played a central role from the very beginning. In particular, the differentiation of field research, which was already carried out in a highly professional manner by the German-French research team Humboldt-Bonpland during the American research journey, but also later during the Russian-Siberian journey with Rose and Ehrenberg, played a decisive role in the practice of *Humboldtian Science.*

For Humboldt's research program on the ground in America or Asia—and the accompanying *collection of* objects, data and information—was ingeniously coordinated with internationally renowned *collections* of artifacts, evidence and information in museums, libraries and archives in France and Prussia as well as in Spain. The fieldwork that took place overseas staked out fundamentally the field of subsequent work in Europe.

Alexander von Humboldt's conception of science—as evidenced precisely by the critical overcoming of an encyclopedic epistemology of the eighteenth century with little empirical foundation—is utterly unimaginable without travel. This applies not only to the great transcontinental journeys to the American hemisphere and Asia but also to the smaller and supposedly 'less important' journeys that took Humboldt to Holland and England, Bohemia or Poland, Austria or Switzerland, Italy or Spain.

Of central importance for Humboldt, however—this should not be overlooked—were the constant movements and journeys between Berlin and Paris, which had begun in 1790 with the spectacular visit to the revolutionary French capital and came to an end with a final stay in Paris between October 1847 and January 1848 on a diplomatic mission for the Prussian king. It was only when Humboldt was physically no longer able that the almost eighty-year-old refrained from travelling to the French capital.

There is no doubt that Paris, with its renowned scientific institutions, its archives, libraries and museums, its publishing houses and scientific circles, but also with its literary and cultural life as well as its salons, was the ideal place for Humboldt as a major European centre to advance his scientific work and to observe that accelerating circulation of knowledge that inspired his work. There is no doubt that the French language, which Humboldt preferred to use for decades, was an ideal tool for him to be perceived on an international stage that was not only national but also extended far beyond Europe. There was undoubtedly no better place in Europe than Paris to survey the multitude of political and social as well as scientific revolutions. And there is no doubt that the forms of science, as well as general sociability that Humboldt enjoyed in Paris, were of inestimable value to the Prussian with regard to the dynamic conception of a world consciousness whose complexity could certainly be perceived far more directly and intensively in Paris than in Berlin. Humboldt knew what he had in his Paris.

And yet the development of Alexander von Humboldt and his life's work was linked less to a specific concrete place than to a vectoricity that emerges in all his writings, for example, in the superimpositions of the most diverse places and landscapes. For Humboldt himself, behind one place there always appeared another, one space always referred to another. Across Humboldt's journeys, which are still astonishing in terms of their dimensions and results, an oscillating movement emerges that continuously—over almost six decades—linked Berlin and Potsdam with Paris. Humboldtian science not only moves between the eighteenth and nineteenth centuries, between the Old and the New World, between the epistemology of the *Encyclopédie* and the epistemology of Darwin.[47] Above all, it also arises from the

[47] See Gould, Stephen Jay: Church, Humboldt, and Darwin: The Tension and Harmony of Art and Science. In: Franklin Kelly et al. (eds.): *Frederic Edwin Church*, Washington D.C.: National Gallery of Art and the Smithsonian Institution Press 1989, pp. 94–107.

constant movements between the German and French languages, between the German- and French-speaking scientific landscapes, between Berlin and Paris.

The two capitals are not so many opposites to be compared with each other, but can rather be understood as points of reference of a wide-ranging relationality, insofar as both cities also embody different circles of friends and scientific networks as well as different lifestyles and styles of thinking.

The name Paris stands not only for a city that fascinated Humboldt throughout his life but also for a crossroads, for a certain vectorial dimension and potentiality that characterize a place like the French capital. Humboldtian thought and scholarship are inconceivable without this *croisement* between Berlin and Paris, Paris and Berlin. It provided Humboldt with the mobility and freedom that were the prerequisites for his democratic science, which was geared towards reaching the broadest possible audience.

When Humboldt 'carting sand for the erection of the temple of freedom' in 1790, he consequently created the conditions for the freedom of his science, his thinking, his own life. He saw the French Revolution and the Declaration of the Rights of Man as the foundation of all coexistence on a world scale, and at the same time, he saw in the revolutions of society and science, as much in Europe as in the Americas, the impetus that would bring about a freer society. The fact that his own life took place for the most part beyond all social (if not scientific) revolutions may be one of the paradoxes of his life and—as the correspondence with Karl August Varnhagen von Ense shows—the adversities of an existence that knew how to come to terms with the circumstances without ever letting go of the ideals and hopes for a better, more open society. So it is not surprising that even in his old age he described himself as a 'tricolored rag'.

Heinrich von Kleist: Saint-Domingue and the Haitian Revolution

From the Breakdown of Living Together

The power, even violence, that Heinrich von Kleist's (1777–1811) texts tend to exude often begins with the title. This is also the case with the narrative text that poses the question of the forms and norms of global coexistence like no other from Kleist's pen. For already with this paratextual element of the 'threshold'[1] begins that peculiar oscillation that pervades the entirety of a narrative that first appeared in the Berlin journal *Der Freimüthige* as a serial story in the issues between 25 March and April 5, 1811 under the title 'Die Verlobung'.

Even the original title raises questions. For this semantically tremendously condensed text is not about an engagement, at least at first glance: an engagement in the proper sense as an announcement of a common will to marry in the future, which—as Roland Reuß has rightly observed—"is unthinkable without the involvement of the public,"[2] does not, as is well known, take place in the entire narrative text. Heinrich von Kleist, who had broken off his own engagement to the general's daughter Wilhelmine von Zenge in 1802 in a brusque manner and from a distance, was certainly aware of this. The text itself underlines this fact quite bluntly[3] and thus draws attention to its own immanent poetics: this is from the very beginning under the sign of a deception and *dissimulation* of meaning.[4]

[1] Cf. Genette, Gérard: *Seuils*. Paris: Seuil 1987.

[2] Reuß, Roland: 'Die Verlobung in St. Domingo'—eine Einführung in Kleists Erzählen. In: *Berliner Kleist-Blätter* (Basel) 1 (1988), p. 6. This essay can be found in the edition of Kleist, Heinrich von, cited below under the sigle V directly in the text: *Die Verlobung in St. Domingo.* Edited by Roland Reuß in collaboration with Peter Staengle. In: Heinrich von Kleist: *Sämtliche Werke. Berlin edition.* Edited by Roland Reuß and Peter Staengle. Vol. II/4. Basel/Frankfurt a. M.: Stroemfeld—Roter Stern 1988.

[3] Cf. Reuß, Roland: *'Die Verlobung in St. Domingo'—an introduction to Kleist's narrative*, p. 6.

[4] The paradoxical unity of life and work of Heinrich von *Kleist as* well as "subtle strategies of dissimulation" was pointed out right at the beginning of his biography by Amann, Wilhelm: *Heinrich von Kleist. Leben—Werk—Wirkung.* Berlin: Suhrkamp 2011, p. 8.

O. Ette, *Mobile Prussia*, https://doi.org/10.1007/978-3-476-05839-3_5

But that was not enough. When the narrative text (after another publication in June as a serialized story in the Viennese journal *Der Sammler*) was first published in book form in August 1811 by the Berlin publishing house of the Realschulbuchhandlung as the opening text of the second volume of Kleist's *Erzählungen*, its author expanded the title turn to its final form: *The Betrothal in St. Domingo*. This seemingly only spatially locating addition, which is commonly pronounced 'in Santo Domingo' in research as well as in the broader public, points to a further deception and misrepresentation of the meaning that nevertheless seems so evident.

For the abbreviation 'St.', which is common in both German and French, does not abbreviate the Spanish 'Santo' (Sto.), which would also make little sense in the context of the narrative, but refers back to the name of the French colony of *Saint-Domingue*, that 'pearl of the Caribbean', which for much of the eighteenth century was considered, with good reason, to be the richest and most profitable colony not only of the 'New', but of the entire world, for its masters as well as for its mother country. Saint-Domingue was the epitome of a colony based on slave economy and slave exploitation par excellence.

The place name added by Kleist is thus a hybrid formation that should be pronounced 'Sankt Domingo', but more appropriately for the diegesis as 'Saint Domingo', a form through which the double affiliation of the entire island in its economically extremely intensively used western part to the French, in its larger and more extensively used eastern part to the Spanish colonial system comes perfectly to the fore. The title, as can already be seen at this point, has it all and builds up an oscillating field of tension.

The island, which was christened 'Española'[5] by Christopher Columbus in late 1492 in a European act of coupling naming and ownership, and which entered the nomenclatures of geographical textbooks and atlases under the designation 'Hispaniola', is consequently a 'double island' still divided between two European colonial powers at the end of the eighteenth century, to which attention is already drawn in the first lines of the narrative. There, of course, the narrator speaks of "the French share of the island of St. Domingo" (V 7), with which the narrator, as always in Kleist's work, correctly points out to his readers the division of the large Antillean island on the one hand, but on the other hand obviously speaks in deliberate deception of an island called "St. Domingo", which—as we have seen—does not exist as such. A mere carelessness or thoughtlessness on Kleist's part?

Hardly. The formulation in the narrative text, identical to the second component of the title, points us to a semantic misalignment that, in the combinatorics of two languages and two traditions of rule, sets in motion a semantic oscillation that first and foremost draws attention to a third, absent thing: to the pre-Columbian name Haiti, which goes back to indigenous roots and reappeared on the political map with the fought-for independence and the triumph of the Haitian Revolution in 1804.

---

[5]On the history of the discovery and colonization of the island of Hispaniola, see Bernecker, Walther L.: *Kleine Geschichte Haitis*. With the collaboration of Sören Brinkmann and Patrick Ernst. Frankfurt a. M.: Suhrkamp 1994, pp. 11–22.

Thus, the slaves deported to the Caribbean by France, the actual leading power of the second phase of accelerated globalization, fell back on the name that the indigenous population exterminated by the Spanish, the dominant leading power of the first phase of accelerated globalization, had once given to their island. The black slaves thus took up the violently suppressed pre-Columbian tradition of the indigenous population originally native to the island, whose failing labour force they replaced, as it were, as 'imported natives' in the Spanish colonial system from the beginning of the sixteenth century. In the hybrid title of this narrative[6] set in the historical context of the Haitian Revolution, both Spain and France are linguistically highly present. In terms of lexis, one could speak of a trilingual title.

In this enigmatic game with the island name, which is a game of deception and dissimulation signalled by the text itself, not only different languages and traditions of meaning are brought together, but also two historical epochs, which we may call the *first* phase of accelerated globalization, reaching its early climax in 1492—in the course of which the Iberian powers Spain and Portugal appropriated large parts of the American continent—and the *second* phase of accelerated globalization, the latter of which was dominated by the colonial powers France and England in the second half of the eighteenth century. The latter phase was dominated by the colonial powers of France and England.[7] As a result of this process, which was not only political and economic but also social and cultural, the Caribbean—and first and foremost the French Saint-Domingue—had become the world's most condensed space of a globalization that emanated from Europe and affected all areas of life and coexistence in this geostrategically central part of the American hemisphere.

In the phrase *St. Domingo* chosen by Kleist, these two phases of European expansion are not only brought together in perhaps the shortest and most condensed way, but also summarized and thought together. It is evident: The title *Die Verlobung in St. Domingo* can be understood as a fractal of titles, insofar as in this formula, which in the reference to the saint and in the emphasis on *St.* at the same time opens up a dimension of transcendence, an immense semantic compression, a historical telescoping, and the immanent poetics of the entire narrative are expressed. Basically, from the beginning of his narrative, Heinrich von Kleist flashes in miniaturized form the entire history of European colonization.

---

[6] On the textual filiation derived from Kleist, cf. Uerlings, Herbert: Die Haitianische Revolution in der deutschen Literatur: H.v. Kleist, A.G.F. Rebmann, A. Seghers, H. Müller. In: *Jahrbuch für Geschichte von Staat, Wirtschaft und Gesellschaft Lateinamerikas* 28 (1991), S. 343–389; as well as Uerlings Herbert: Preussen in Haiti? On the intercultural encounter in Kleist's 'Betrothal in St. Domingo'. In: *Kleist-Jahrbuch* (1991), pp. 185–201.

[7] Cf. for a first elaboration of this periodization Ette, Ottmar: *Weltbewußtsein. Alexander von Humboldt und das unvollendete Projekt einer anderen Moderne.* Weilerswist: Velbrück Wissenschaft 2002, S. 26 f.; as well as Otmar Ette: Wege des Wissens. Five Theses on World Consciousness and the Literatures of the World. In: Hofmann, Sabine/Wehrheim, Monika (eds.): *Lateinamerika. Places and Orders of Knowledge. Festschrift for Birgit Scharlau.* Tübingen: Gunter Narr 2004, pp. 169–184.

This fact is already pointed out by the observed close interlocking of title and *incipit* of Kleist's narrative text, as expressed in the opening sentences, which cannot be called anything other than spectacular:

> At Port au Prince, on the French part of the island of St. Domingo, at the beginning of this century, when the blacks were murdering the whites, there lived on the plantation of Mr. Guillaume of Villeneuve a terrible old negro named Congo Hoango. This man, who came from the Gold Coast of Africa, and who in his youth seemed to have been of a faithful and upright disposition, had been showered with infinite favors by his master, because he had once saved his life on a passage to Cuba. Not only did Mr. Guillaume immediately give him his freedom, and, on his return to St. Domingo, gave him his house and farm; he even made him, a few years later, against the custom of the country, overseer of his considerable estate, and, because he did not wish to marry again, placed with him in his wife's place an old mulatto woman, named Babekan, from his plantation, to whom he was distantly related through his first deceased wife. Yes, when the negro had reached his sixtieth year, he retired him with a handsome salary, and crowned his benefactions by even making him a legacy in his legacy; and yet all these proofs of gratitude could not protect Mr. Villeneuve from the fury of this fierce man. (V 7 f)

The very first sentence of Kleist's master narrative inserts a spatial and temporal situating and contextualization in which—unusually for the work of the author born in Frankfurt an der Oder—the events of the Haitian Revolution, only a few years in the past, serve as an immediate historical and geographical frame of reference. In this *incipit*, the narrator's voice—which we must not confuse with the real author outside the text, as can be observed time and again—takes an unmistakably partisan position insofar as the uprising and revolution against the plantation system based on slavery are staged as a space of time and conflict in which simply "the blacks murdered the whites" (V 7).[8] This initial black-and-white image, clearly marked as such, thus presents the complex historical situation as that of a race war in which the whites became the victims of the blacks.

There is no doubt about it: the historical background can be described as the greatest presumable accident of coexistence, whereby it is quite natural to assume that the previously prevailing state of conviviality was based on extreme colonial exploitation and brutal oppression, as characterized the European systems of rule since the beginning of the first phase of globalization. Revolutionary France reacted to the no less revolutionary developments in its colony with constantly new changes of course and decrees. When Napoleon, who had come to power, finally attempted to reconquer the rich colony for France and reintroduce slavery by sending an expeditionary army commanded by General Leclerc, which landed in Saint-Domingue on February 1, 1802, with the most massive deployment of French troops, the

[8] On the question of Kleist's racism, cf. Loster-Schneider, Gudrun: Toni, Babekan und Homi Bhabha? On Problems of Cultural and Aesthetic Hybridization in Heinrich von Kleist's 'The Betrothal in St. Domingo'. In: Lüsebrink, Hans-Jürgen (ed.): *Das Europa der Aufklärung und die außerereuropäische koloniale Welt.* Göttingen. Wallstein 2006, p. 231; there are also numerous further bibliographical references.

revolutionaries stubbornly and resolutely put up a fight: the final phase of the future Haiti's struggle for independence, rich in surprising twists and turns, began.[9]

It is not without reason that the formulations of the *incipit* make no mention whatsoever of the cruelty of the Napoleonic troops, who ruthlessly proceeded against the insurgents and mercilessly massacred them. In the contemporary European press, the 'Negro uprising', which was to go down in world history as the first successful slave revolution and form the first politically independent country of the emerging 'Latin America'—this designation has only been used since the middle of the nineteenth century[10]—appeared as a bloodbath perpetrated against the whites, indeed as a dangerously flaring up 'world fire'. The 'Negro slaves of Haiti' had, as one might put it with Hans Christoph Buch, 'taken Robespierre at his word'[11] and fought for their own independence under the sign of hoped-for *liberté*, *egalité* and *fraternité.* Like many of his contemporaries, Heinrich von Kleist was—as will be explained in more detail—of course very familiar with the events on the island of Hispaniola.

Colonial Island Worlds

Even though the Haitian Revolution hardly ever found its way into the development of Western revolutionary theories for a long time,[12] right up to the present day, the events from Hispaniola, which were often dismissed as an 'outrage', quickly became a paradigm[13] and an exceptional case of a catastrophe that triggered fear and terror

[9] Cf. also Nesbitt, Nick: The Haitian Revolution and the Globalization of the Radical Enlightenment. In: Ette, Ottmar (ed.): *Caribbean(s) on the Move—Archipiélagos literarios del Caribe. A TransArea Symposium.* Frankfurt a. M./New York/Oxford: Peter Lang 2008, pp. 40–59.

[10] Cf. Jurt, Joseph: Entstehung und Entwicklung der Lateinamerika-Idee. In: *Lendemains* (Cologne) 27 (1982), pp. 17–26.

[11] Buch, Hans Christoph: *Die Scheidung von San Domingo. Wie die Negersklaven von Haiti Robespierre beim Wort nahmen.* Berlin: Klaus Wagenbach 1976.

[12] Cf. on this Müller, Gesine: *Die koloniale Karibik zwischen Bipolarität und Multirelationalität. Transfer Processes in Hispanophone and Francophone Literatures.* Habilitationsschrift an der Philosophischen Fakultät der Universität Potsdam, Potsdam 2011, p. 128 f. The 'case of Haiti' is more than problematic with regard to the prevailing Western theory of revolution: in her foundational work *On Revolution* (New York: The Viking Press 1963), Hannah Arendt herself refrained from mentioning Haiti. For a long time, colonialism and slavery have been regarded in the theory of modernity—and especially in the theory of revolution—merely as marginal phenomena or even as disruptive factors (see Fischer, Sibylle: *Modernity Disavowed. Haiti and the Cultures of Slavery in the Age of Revolution.* Durham/London: Duke University Press 2004, p. 8 f.). A scholarly turn to the Haitian Revolution has recently become unmistakable; cf. on this, among others, Bongie, Chris: *Friends and Enemies. The Scribal Politics of Post/Colonial Literature.* Liverpool: Liverpool University Press 2008.

[13] Cf. inter alia Bremer, Thomas: Haiti como paradigma. La emancipación de los esclavos en el Caribe y la literatura europea. In: López de Abiada, J.M./Peñate Rivero, J. (eds.): *Perspectivas de comprensión y explicación de la narrativa latinoamericana. Grandes Seminarios de Travers.* Bellinzona: Ed. Casagrande 1982, pp. 43–66. The extraordinary presence of the events on Saint-Domingue as well as the independence of Haiti, especially in the German-speaking world around

among the whites not only in the Caribbean region, but also in the colonial mother countries of Europe. The "birth of Haiti"[14] was an extremely complex, but at the same time necessarily violent process, which was depicted in Europe, but also in the neighbouring colonies of the Caribbean in blood-red colours, since the white ruling elites had to fear that, similar to Saint-Domingue, they would be swept away by a numerically superior population of black slaves and at best forced into exile.

Against this background, it seems logical for Kleist's narrator to presuppose such a view as a widespread opinion, as it were, in order to draw from it a horrifying picture of an abruptly imploding, collapsing coexistence. *The Betrothal in St. Domingo* narrates from great temporal proximity the collapse of conviviality in the colonial context, which—as emphasized—was based on the rules of a mercilessly functioning caste system. The lives, experiences, and hoped-for survival of a small group of whites trying to make their way to Port-au-Prince are at the forefront of the literary account. Let us turn, then, to the decisive plot elements of Kleist's narrative.

Thus the very first verb of the narrative ("lived") signals the decisive level of meaning of life and conviviality, which has also clearly been incorporated into the highly thought-out title. For what else could an engagement be but a formalized promise of a future life together? The prospective dimension of the lexeme engagement as well as the perspective of the future will be returned to repeatedly in this context.

So if the isotopia of life and conviviality dominates the title and prelude of *The Betrothal in St. Domingo in* an almost unmistakable—and yet generally overlooked—way, this does not happen without the relations of domination between the masters and the slaves also being presented in all their impressiveness. Here, too, the narrator makes use of a narrative artifice, insofar as the assessment of colonial hierarchy and asymmetry to be expected from a European reading public is included in the text as a 'self-evident' setting, as it were. For Guillaume *von* Villeneuve—here, too, we are dealing with a bilingual naming convention—is dubbed Lord in multiple repetitions, while a "terrible old negro, named Congo Hoango" (V 7) has no claim whatsoever to a respectful form of address and titling. The macro-history is linguistically translated down to the last ramifications of the narrative.

In the case of Congo Hoango, too, Kleist's choice of name is revealing, since this black man is not a slave born on Hispaniola, but a 'man' deported from Africa, in whose name two rivers, the Congo and the Huang-he, i.e. an African and an Asian river, flow together in a double name. In doing so, the Chinese name for the 'Yellow River' covertly—though recognizably[15] to users of contemporary reference works—blends the color 'yellow' into the text for the first time. The black called up by the name 'Congo' is subtly joined by the yellow as a third, which is placed between the

---

the journal *Minerva, was* pointed out in detail by Buck-Morss, Susan: Hegel and Haiti. In: *Critical Inquiry* (Chicago) 26 (Summer 2000), pp. 821–865.

[14] Bernecker, Walther L.: *Kleine Geschichte Haitis*, p. 37.

[15] Cf. the quotation from Zedler's *Großer Vollständiger Universallexicon aller Wissenschaften und* Künste. In: Reuß, Roland: 'Die Verlobung in St. Domingo'—eine Einführung in Kleists Erzählen, p. 18.

white and the black in the play of colours soon explicitly set in motion by Kleist's narrative.

Congo Hoango is subsequently introduced to us first as the life-saver of his French owner and immediately afterwards as the murderer of the same Lord of Villeneuve. The fact that Kleist has the master's rescue by his slave take place during a crossing to Cuba, of all places, i.e. precisely that crossing between the two Greater Antilles which anticipates the later escape route of the planter aristocracy of Saint-Domingues to the neighboring island, is by no means due to chance. For this vectorial or inscription of the future exodus of the representatives of a swept-away ruling structure translates the life-saving action of the black man a little later into a death-dealing action, in that Villeneuve and his family are no longer given the chance to make the saving escape to the neighboring island of Cuba by the same route.

The archipelagic multi-connectedness of the colonial Caribbean already appears clearly at the level of the characters in Kleist's narrative: the mulatto Babekan is fathered by a white man in Santiago de Cuba (v 20); her daughter Toni is born in Europe; Congo Hoango was deported from Africa; Villeneuve comes from France; Gustav or Strömli come from Switzerland—the list could easily be continued. This multi-connectedness, however, is interrupted in the Haitian Revolution with far-reaching consequences, not only economically: The *world of islands* (Inselwelt) tends to become an *island as a world of its own* (Insel-Welt).[16] Kleist's narrator presents the 'island' of St. Domingo as such.

Even if they are presented affirmatively by the narrator, the "benevolences" of the Lord of Villeneuve are nevertheless proofs of favor from a slave owner who sought to give a more human face to the system of an extremely profit-oriented slave economy for which the French sugar colony of Saint-Domingue stood paradigmatically. It was no coincidence that it was the French sugar oligarchs from Saint-Domingue who, after fleeing to the neighboring island, soon transformed the Spanish-owned Cuba into the world's largest sugar producer, not without being instrumental in ensuring that the 'import' of black slaves for the cultivation of sugar cane and coffee multiplied within a few decades and that new techniques of exploitation—which can probably only be described as 'modernization' under the sign of a cynical rationality—were implemented.[17] Thus, the *know-how of* the leading power of the second phase of accelerated globalization was transferred to the colonial empire of Spain, a fact without which the exceptional position of Cuba in the nineteenth century—as much with regard to the observable economic rise as to the island's remaining in Spanish colonial possession—can hardly be adequately understood.

---

[16] Cf. Ette, Ottmar: Von Inseln, Grenzen und Vektoren. Versuch über die fraktale Inselwelt der Karibik. In: Braig, Marianne/Ette, Ottmar/Ingenschay, Dieter/Maihold, Günther (eds.): *Grenzen der Macht—Macht der Grenzen. Latin America in a Global Context.* Frankfurt a. M.: Vervuert 2005, pp. 135–180.

[17] On the development of the slave economy and sugar production at the turn of the eighteenth and nineteenth centuries in Cuba, see the classic study by Moreno Fraginals, Manuel: *El Ingenio. El complejo económico social cubano del azúcar.* Vol. 1: 1760–1860. La Habana: Comisión Nacional Cubana de la UNESCO 1964.

Therefore, the crossing to Cuba inserted by Kleist anticipates a historical development whose consequences were already clearly foreseeable in 1811, a few years after Haiti's independence. The release of Congo Hoango interwoven into the narrative simultaneously transformed the black slave into an overseer at the behest of his master, who of course also had to 'supervise' his master's slaves, tasks that thus placed him in a position between his master and the slaves subordinate to him, whereby the connection created by Villeneuve with the mulatto Babekan additionally translated this intermediate position into a socio-culturally anchored scale of skin colors between black and white. Congo Hoango is not a simple black slave: as *dominé dominant in* the colonial system of Saint-Domingue, he is used to issuing orders and commands himself.

The freed slave, however, did not thank his master's gratitude, but—in the sequence of all those political events that had only been triggered by the "rash steps of the National Convention"—"put a bullet through his master's head", mind you: "mindful of the tyranny that had snatched him from his fatherland" (v 8). The master is judged in this sense as representative of a system of transatlantic slave trade, as tyrannical as it was lucrative, which had developed very much in the context of the second phase of accelerated globalization, and had even 'modernized' and thus taken to extremes the most intense and inhuman forms of exploitation of man by man. Villeneuve dies as the representative of a French colonial system operating worldwide, which was only surpassed in contemporary terms by the British *Empire.*

Let us not forget at this point how much Heinrich von Kleist, in a letter to Rühle von Lilienstern in December 1805, had been outraged by how "out of the whole cultivated part of Europe a single great system of empires"[18] was in the process of being created, which was in complete dependence under the leadership of Napoleon—who put on the imperial crown himself in December 1804. And not for nothing did this letter end with the hope of a tyrannicide in phrases that come very close to the talk of the tyranny of the slave trade in *The Betrothal at St. Domingo*: "Why only One cannot be found to put the bullet through the head of this evil spirit of the world."[19]

Against the background of this system, which he experienced in all its brutality, Congo Hoango could no longer think of peaceful coexistence with the whites:

> [...] he set fire to the house in which his wife had taken refuge with her three children and the other whites of the settlement, devastated the entire plantation to which the heirs who lived in Port au Prince could have laid claim, and, when all the establishments belonging to the estate had been levelled with the earth, went about the neighborhood with the negroes whom he had assembled and armed in order to assist his confreres in the fight against the whites. (V 8 f)

[18] Kleist, Heinrich von: *Sämtliche Werke und Briefe in vier Volumes.* Edited by Ilse-Marie Barth, Klaus Müller-Salget, Stefan Ormanus and Hinrich C. Seeba. Vol. IV. Frankfurt a. M.: Deutscher Klassiker Verlag 1987 ff, p. 352.

[19] Ibid.

Thus, for example, in the talk of tyranny in and under the narrator's voice in *Die Verlobung in St. Domingo*, other voices become audible from the very first pages, reporting not only on the murder of whites by blacks, but also on the tyranny of a global economic system of sugar cane plantations based on slavery. Among these contemporary voices was a well-travelled Prussian compatriot of Heinrich von Kleist, who, unlike the latter, was not hostile to France, but expressed profound doubts about the increasingly ruthlessly 'modernized' colonial system and its slave economy.

Haiti as a Challenge and Paradigm

Heinrich von Kleist was certainly familiar with aspects of the so-called 'dispute about the New World',[20] which originated in Cornelius de Pauw's *Recherches philosophiques sur les Américains* published in Berlin in 1768 and 1769 and reached its first climax with the 'Berlin Debate', which was ignited by Antoine-Joseph Pernety's lecture of September 7, 1769, held at the Berlin Academy and directed against de Pauw.[21] Even in Hegel's Lectures on World History, long decades later, the disputes that had been fought out for decades not only between European philosophers, but precisely between New-World and Old-World philosophers, still resonate. Clear traces of these sometimes heated debates and their results can also be found in *Die Verlobung* in *St. Domingo.*

More explicitly, however, for diegetic reasons, the political debates in France are outlined in this narrative text. The narrator's aforementioned reference to the "rash steps of the National Convention" (V 8) points beyond the singular debates and decrees of the various constitutional bodies created by the French Revolution to the political disputes in Paris, which were of the utmost importance for the administrative framework of a conviviality of the most diverse social and ethnic groups in the French colonies.

Not only the great colonial debate conducted from 7 to 15 May 1791 in the *Constituante*, but above all the abolition of slavery in the colonies decided by the National Convention on February 4, 1794 led to immediate reactions, also and especially in the French Caribbean. These reactions illuminate the speed of political as well as administrative implementation that characterized transatlantic relations in the second phase of accelerated globalization. It is no coincidence that the Caribbean, as a zone of the most concentrated globalization, had very rapid connections that

[20] Cf. the standard work by Gerbi, Antonello: *La Disputa del Nuovo Mondo. Storia di una Polemica: 1750–1900.* Nuova edizione a cura di Sandro Gerbi. Milano/Napoli: Riccardo Ricciardi Editore 1983. Cf. on this also the second and especially the third chapter in the present book.

[21] Cf. the text by Pernety, Antoine-Joseph: Dissertation sur l'Amérique et les Naturels de cette partie du monde. In: Mr. de P*** [Pauw, Cornelius de]: *Recherches philosophiques sur les Américains ou Mémoires intéressants pour servir à l'Histoire de l'Espèce Humaine.* Avec une Dissertation sur l'Amérique & les Américains, par Dom Pernety. Et la Défense de l'Auteur des Recherches contre cette Dissertation. Berlin: Chez Georges Jacques Decker, Imp. du Roi 1770.

kept the colonies in close exchange with their respective European mother countries across the Atlantic longitudinal valley.

Comprehensive analyses of the demographic foundations and the economic effects of these processes were only achieved with a considerable time lag in studies that turned to the entirety of the region we now call the *Greater Caribbean* from a globalizing perspective. In his *Essai politique sur l'île de Cuba*, published separately for the first time in 1826 and written in the context of his French-language travelogue *on the Voyage aux Régions équinoxiales du Nouveau Continent*[22] undertaken together with Aimé Bonpland, Alexander von Humboldt repeatedly drew attention to the extremely inhumane production conditions on Saint-Domingue, underpinning his reflections with extensive statistical material.[23]

The Prussian writer and scholar, who at the time of the Haitian Revolution twice spent several months on the neighboring island of Cuba, had of course already dealt much earlier with the slave economy, which in his scientifically founded view was detestable and unjustifiable, from the perspective of the human rights proclaimed in Paris. Regarding the treatment of slaves, he noted on December 6, 1800, aboard a ship bound for Cuba: "Nothing can be exaggerated, so shameful is this treatment."[24] In Europe, he said, people speak admiringly of the splendor of Saint-Domingue, but everything unnatural must sooner or later disappear: "and it is not natural that a few rocky islands should produce so much."[25] Humboldt was aware that the generation of such great wealth could only be possible by exploiting slaves to their utmost limits.

In his *Essai politique*, he repeatedly demonstrated, using concrete figures, the extent to which Saint-Domingue had taken the system of plantation economy to extremes. And had not the collapse of the French colony also triggered the increased importation of black slaves into Cuba? Humboldt proved this point by looking at the official import figures for the port of Havana, showing two great leaps in his coverage of the years 1790 to 1820. Between 1801 and 1802 the number of imported slaves rose from 1659 to 13,832, and between 1814 and 1816 from 4321 to 9111 to

[22] Cf. Humboldt, Alexander von: *Relation historique du Voyage aux Régions équinoxiales du Nouveau Continent …* Reprint of the complete original published in Paris 1814–1825, procured, introduced and increased by an index by Hanno Beck. 3 vols. Stuttgart: Brockhaus 1970.

[23] Cf. on this the details and numerous notes also in the recently published English-language edition of this fundamental work in Humboldt, Alexander von: *Political Essay on the Island of Cuba. A Critical Edition.* Edited with an Introduction by Vera M. Kutzinski and Ottmar Ette. Translated by J. Bradford Anderson, Vera M. Kutzinski, and Anja Becker. With Annotations by Tobias Kraft, Anja Becker, and Giorleny D. Altamirano Rayo. Chicago/London: The University of Chicago Press 2011. Corresponding passages can easily be found via the index. For a very first version of Humboldt's essay on the island of Cuba with views of the entire Caribbean, see Ette, Ottmar: InselText und archipelisches Schreiben: Alexander von Humboldt's 'Isle de Cube. Antilles en général'. In: *edition humboldt digital.* Ed. By Ottmar Ette. Berlin: Berlin-Brandenburg Academy of Sciences and Humanities. Version 1 of 10.05.2017 http://edition-humboldt.de/v1/H0016213

[24] Humboldt, Alexander von: *Latin America on the eve of the independence revolution. an anthology of impressions and judgments from his travel diaries.* With an introductory study by Manfred Kossok. 2nd, revised and improved edition. Berlin: Akademie-Verlag 2003, p. 247. I thank Tobias Kraft for pointing out this passage.

[25] Ibid., p. 248.

17,737, reaching a peak of 25,841 slaves deported from Africa through the port of Havana alone in 1817.[26]

What slavery and the plantation economy meant for the coexistence of people in dependent colonial systems, Humboldt himself had repeatedly experienced at first hand. On the crossing from Guyaquil to Acapulco, for example, he was horrified to learn that a fellow passenger, a well-dressed young man, had murdered one of his female slaves with a knife on board the ship. In his diary entry of March 4, 1803, Humboldt not only expressed his deep disgust, but also placed this scene in the context of current political developments, with which Napoleonic France—not having been taught any better by the bloody uprisings against the tyranny of the slave economy—wanted to bring the colonies, and Saint-Domingue in particular, back under its control by reintroducing slavery:

> This new situation has caused us even more distress, because a few days before our arrival, the newspaper had published Bruix's appalling speech, in which, as Government spokesman, he proposed to the Legislative Council that it should reinstate the law and the practice of the Nègres in the same way as it had existed before 1789. We were fortunately in a society in which many people shared the moral horror with which one must receive such distressing news. Here is the fruit of so much blood spilled in the Indies […].[27]

Early on, Humboldt grappled with the economic, political, legal and social consequences that the bloody developments of Saint-Domingue would have for the worldwide network of the plantation economy. For the consequences of the (reintroduction of) slavery were devastating.

Thus, in a letter dated June 20, 1804, from Philadelphia to William Thornton, he sharply criticized the slave legislation in the southern states of the United States and combined his criticism with the vehement question of justice as well as the containment of a purely economic way of thinking:

> The more the recent events in Santo Domingo have offended the truth, the more it seems to be the duty of every moral man to place the problem in its true light. […] How beautiful it would be if the southern states prepared the thing without waiting for the dangerous crisis of 1808. This abominable law which permits the importation of Negroes into South Carolina is a disgrace to a State, in which I know there are very well organized heads. If we follow the only course dictated by humanity, we shall doubtless export less Cotton in the beginning. But alas, how I hate this Policy which measures and evaluates public happiness simply according to the value of the Exports! […] Before being free, we must be fair, and without justice there can be no lasting prosperity.[28]

With greater temporal distance, Alexander von Humboldt warned in his analysis of Cuba's situation in the Caribbean context against a repetition of the end of the once so rich French colony of Saint-Domingue. In view of a steadily increasing proportion of black slaves in the total population, he emphasized the increasingly likely

[26] Humboldt, Alexander von: *Relation historique*, vol. III, p. 404.

[27] Ibid., p. 249 f.

[28] Humboldt, Alexander von: *Briefe aus Amerika 1799–1804.* Edited by Ulrike Moheit. Berlin: Akademie-Verlag 1993, p. 300.

possibility that, without any intervention on the part of the neighboring Haitian Republic, a new revolutionary situation could also arise on other Antillean islands, which would have to lead to a profoundly changed geopolitical situation:

> If the jurisprudence of the Antilles and the condition of the colored population do not soon undergo helpful changes, if people continue to discuss without acting, the political preponderance will fall into the hands of those who are in possession of the labour, the will to free themselves, and the courage to endure long privations. This bloody catastrophe will occur as a necessary consequence of circumstances, and without the free blacks of Haiti interfering in the least, that is, without abandoning the system of isolation which they have hitherto followed. Who would dare to predict the influence which an African Confederation of the Free States of the Antilles, situated between Colombia, North America, and Guatemala, would exert on the politics of the New World?[29]

By raising the possibility of the emergence of an 'African Confederation' in the centre of the American hemisphere, Alexander von Humboldt outlined the radical openness of a history that would no longer follow any predetermined pattern in its course. For the former slaves of Haiti had long since proved that they were capable of unleashing a political *agency* of the greatest magnitude, an ability that Humboldt henceforth took into account. In the wake of the French Revolution, the *Historia* had ceased to serve as a *magistra vitae* in the "horizon of modern history":[30] The objects of European colonial policy had begun to become subjects of their own history. And they were not willing to let the reins of action be taken out of their hands once again.

This raised the question of the future and, above all, of the possibilities and limits of living together within a fundamentally changed constellation of power in the New World: within a new distribution of violence and influence in which the former slaves—following the example of the blacks in Saint-Domingue—could become masters of their own history, their own destiny. But how then would the foundations and 'rules of the game' of coexistence change at a time when slavery continued to prevail in many regions of the circum-Caribbean area, including the southern states of the USA?

---

[29] Humboldt, Alexander von: *Relation historique du Voyage aux Régions équinoxiales du Nouveau Continent …* Reprint of the complete original published in Paris 1814–1825, procured, introduced and augmented by an index by Hanno Beck. Vol. III. Stuttgart: Brockhaus 1970, p. 389: "Si la législation des Antilles et l'état des gens de couleur n'éprouvent pas bientôt des changemens salutaires, si l'on continue à discuter sans agir, la prépondérance politique passera entre les mains de ceux qui ont la force du travail, la volonté de s'affranchir et le courage d'endurer de longues privations. This catastrophe has taken place as a necessary consequence of the circumstances, and without the free blacks of Haiti being aware of it, without them abandoning the system of isolation that they have followed until now. Who is to predict the influence that the *Confederation of African Free States of the Antilles*, placed between Colombia, the Americas and Guatemala, will exert on the politics of the New World?"

[30] Cf. Koselleck, Reinhart: Historia Magistra Vitae. On the dissolution of the topos in the horizon of modern history in motion. In Reinhart Koselleck: *Vergangene Zukunft. On the Semantics of Historical Times*. Frankfurt a. M.: Suhrkamp 1984, pp. 38–66.

Even in the event that Saint-Domingue—contrary to the fears expressed by colonial masters and sugar cane planters throughout the Caribbean—did not interfere in the affairs of the European colonial powers[31] or the USA as an ultimately only small political power, the question was now posed in a radical way as to how a future coexistence was to be shaped and organized on a national and regional as well as on a hemispheric and transareal scale. The narrative examined here is also situated within this transareal discursive field.

For one may claim without any exaggeration that Heinrich von Kleist's *Die Verlobung in St. Domingo* reacts as if in a literary burning mirror to all those challenges and attempts to provide complex answers that Alexander von Humboldt's reactions to the Haitian Revolution present in an almost exemplary manner. This does not mean any kind of 'realistic' mimesis, no 'portrayed reality' in the traditional sense, but a condensed representation of *lived* and *livable* reality, in which the historically *found* is configured with the literarily *invented* into a testing space in such a way that what *can be lived* and experienced can not only be imagined in this first triangle, as in a laboratory situation, but at the same time explored and, even more, *acted out*, even *experienced by* the reading public. *Die Verlobung in St. Domingo* forms, it seems to me, in an almost ideal way, such an experimental testing space—and not at all coincidentally on the example of that Haiti which, as a challenge and paradigm on a global scale, at the same time radically posed the question of conviviality on a worldwide scale.

Precisely because Kleist's narrative also incorporates stereotypical depictions of a battle of the races in Saint-Domingue into its literary modeling from the very beginning, it becomes possible to understand this narrative as a testing ground for the opportunities and risks of future coexistence—a coexistence not only on Hispaniola or in the Caribbean, but in the context of the second phase of accelerated globalization throughout both the New and the Old World. *Die Verlobung in St. Domingo* thus advances to a highly ambiguous and polyphonic text that responds to the challenges of a practice of globalization deeply shaken by the Haitian Revolution and develops complex offers of meaning.

These offers of meaning pose the question of how, at the beginning of a new historical epoch, on whose horizon not only the demise of a plantation economy based on slavery, but of the entire colonial system of European character, can new forms and norms of life be developed that would be able to regulate a future peaceful coexistence. But how could *Die Verlobung in St. Domingo* be understood and read beyond the treatment of the historical paradigm in a prospective sense as a testing ground for future coexistence?

[31] From an admittedly very specific perspective, Kleist's narrative is analysed in Zantop, Susanne: *Colonial Fantasies. Conquest, Family, and Nation in Precolonial Germany, 1770–1870*. Durham/London: Duke University Press 1997, p. 155.

Figural Constructions

Let us ask at this point about the conditions under which the text was written. In Kleist research it is undisputed that Heinrich von Kleist may have received decisive inspiration for this material through his stays in the years 1801/02 as well as 1803 in Paris, 1802 in Switzerland and above all through his imprisonment as a French prisoner of war in 1807 in the imposingly situated fortress of Joux in the Jura not far from the Swiss border.[32] A memorial stone still commemorates the fact that François Dominique Toussaint-L'Ouverture (1745–1803), the once powerful 'slave general' and 'black Spartacus', who was betrayed to the Napoleonic troops on June 15, 1802 and taken to France under inhumane conditions, perished miserably here in a narrow cold cell—and this in the very death cell in which Kleist's acquaintance Gauvain was imprisoned four years later.[33] Kleist was probably supplied with important material not only through the French and German press, through his stays in Paris during the final phase of the Haitian Revolution, and through his friend Zschokke, who was well informed about the events in Saint-Domingue, but also through his almost close contact with the world of Toussaint-L'Ouverture,[34] which must have been so distant from Kleist and yet so fascinating to him.

But was that world really so far away? Had not the new global order that Napoleon devised for a world under French domination washed the Haitian slave general Toussaint-L'Ouverture as well as Kleist, who came from an old line of Prussian generals, into the Forteresse de Joux? Were not the occupation of rebellious Haiti and the occupation of Berlin by French troops directly connected? And was Prussia not about to degenerate into a mere inner-European colony, a puppet in France's global power game?

Some features of the figure of the bloodthirsty Congo Hoango may recall character traits that were attributed to Toussaint-L'Ouverture and—especially with regard to an "inhuman vindictiveness" (V 9)—even more to Dessalines in colonialist discourse. In any case, the curtain of Kleist's narrative is deliberately raised only during the absence of Congo Hoango, who, as we have seen, is no ordinary black man but possesses a certain power and command. As a representative of the emerging new ruling structures, he is a black lord who, unlike his retainers who live in the stable, is able to inhabit "the main building of the plantation" (V 9) with old Babekan and her daughter, "a young fifteen-year-old mestizo" named Toni.

This lonely building on a country road becomes the actual setting for a story whose threads cross and knot in this house. For the house, located somewhere between Fort Dauphin and Port-au-Prince, becomes the central protagonist of the narrative, as *secret* as it is sinister.

---

[32] Cf. Loster-Schneider, Gudrun: *Toni, Babekan and Homi Bhabha? On Problems of Cultural and Aesthetic Hybridization in Heinrich von Kleist's 'Die Verlobung in St. Domingo'*, p. 228.

[33] Ibid., p. 230. Compare also Kleist, Heinrich von: *Sämtliche Werke in vier Volumes*, Vol. IV, p. 374.

[34] Cf. also Bernecker, Walther L.: *Kleine Geschichte Haitis*, esp. p. 41. A short biographical sketch of the revolutionary general, who was born a slave, can also be found in Buch, Hans Christoph: *Die Scheidung von San Domingo*, p. 173 et seq.

This house, which until recently had belonged to the slave owner Guillaume of Villeneuve, initially appears as a trap, as Congo Hoango and Babekan successfully try to lure whites in need of help into the house with the active support of the beautiful Toni, "who, because of her yellowish complexion, was particularly useful for this ghastly ruse" (V 10). A second triangle opens up: between a murderous *cunning* that allows a *burden* inherited through slavery and colonialism, consisting of humiliations and abuses repeatedly named and suffered until recently, to be transformed into a *lust* that, as murderous lust, is undoubtedly in the sign of revenge against the whites and their cruel system of oppression.

Kleist's narrative presents this lust of Congo Hoango as well as of Babekan as reprehensible from the very beginning, since it is placed on the surface of the text under the sign of an ingratitude that transformed the murderer of the benevolent master into a master of cruel murderers. The fair-skinned Toni accepts this approvingly and has already lured Portuguese or Dutchmen into the deadly trap of the house besides several Frenchmen.

In the narrative, not only blacks and whites confront each other. Between white and black stand the mulatto woman Babekan and her daughter Toni, who—as her yellowish complexion indicates—is also mulatto. The text refers to her conceptually incorrectly as a "mestizo", since Toni in no way emerged from a connection between the indigenous (Indian) population and white immigrants. On a structural level, however, this designation is revealing not only because Heinrich von Kleist may have followed a contemporary turn of phrase in the French Antilles, in which the designation as *métisse* tends, as it were, to omit the 'black', African part in order to valorise the corresponding person in terms of his white origins.[35] For the lexeme "mestizo", which Kleist obviously chose very consciously for Toni, integrates additional levels of meaning; in this way, the young woman in her iridescent form, in her fundamental "in-between-ness", blends in the indigenous population, which had already been largely exterminated by the Spaniards, so that the beautiful Toni moves on the conceptual level in the third triangle between Europe, Africa and America. The "mestizo" is a figure of movement on the most diverse levels.

The narrator's ostentatious choice of seemingly 'erroneous', 'inaccurate' terminology transforms Toni into a being who, more than any other person in Kleist's narrative, is able to stand up for the Caribbean in both the first and second phases of accelerated globalization. The fact that her father is a rich French merchant from Marseille (V 27) who later went to the Turkish court in Istanbul as an envoy of the young French Republic (V 28) reinforces this aspect: Toni is a product of all those movements in which the East Indies and West Indies, Asia, America and Africa were globalized from Europe.

[35] In various contexts, this conceptual usage is illuminated by the well-traveled writer of Irish-Greek descent Lafcadio Hearn; cf. his original travelogue *Two Years in the French West Indies*, less noted than the Japanese writings, reprinted in Hearn, Lafcadio: *American Writings*. New York: The Library of America 2009, pp. 149–541.

Certainly, the color yellow can be located in the color spectrum between black and white, as has been elaborated again and again in research.[36] However, yellow also inscribes itself in a long patriarchal-colonialist and sexist tradition in the Caribbean, in which from the perspective of the white man—and here another, fourth triangle opens up, this time concerning dependent gender relations—the black woman was primarily intended for work, the white woman primarily for marriage, and the mulatto woman for lust and love.[37]

No wonder, then, that the beautiful mulatto became not only a myth, but soon also the subject of literary allegories that referred to the emerging nations of the American hemisphere. For in the further course of the nineteenth century in the Spanish-speaking world—and the neighboring island of Cuba certainly offers the best example here—the *mulata* advanced to become the embodiment of the future nation, for her allegory, in the context of problematic nation-building processes, offered the possibility of demonstrating both the beauty and the failure of this project of aspired conviviality in a way that was as sensuous as it was skin-deep, with a high identificational value.[38] In terms of motifs and material history, Kleist falls back on the mulatto woman as a central figure at an early stage.

In the French-speaking Caribbean, the color yellow had established itself far more strongly and far earlier than in the Spanish-speaking world as a euphemism for everything 'mulatto'. Thus, this racially tinged 'color theory' is found in relevant texts such as Moreau de Saint-Méry, who refers to yellow as a third shade alongside white and black: "The third shade is that of the Mulatto, which could almost be subdivided into two, since Mulattos compared to each other offer two very distinct shades which are exactly that of red copper and that of yellow copper." ("La troisième nuance est celle du Mulâtre qu'on pourrait presque subdiviser en deux, attendu que les Mulâtres comparés entr'eux, offrent deux nuances très distinctes qui sont exactement celle du cuivre rouge & celle du cuivre jaune.")[39] Médéric-Louis-Elie Moreau de Saint-Méry, in his 1802 work on Saint-Domingue, which became an important source for Heinrich von Kleist, had brought into the field[40] highly complicated color subdivisions, as, moreover, can be found

[36] Cf. for example Uerlings, Herbert: *Poetiken der Interkulturalität. Haiti bei Kleist, Seghers, Müller, Buch und Fichte.* Tübingen: Max Niemeyer 1997, pp. 25–40.

[37] Cf. on this, among others, Phaf, Ineke: Motivforschung altmodisch? With the Mulatto to a Caribbean National Text. In: Kohut, Karl (ed.): *Race, Class and Culture in the Caribbean. Proceedings of the symposium 'Race Relations and Race Encounters in the Caribbean', May 15 and 16, 1987, at the Catholic University of Eichstätt.* Frankfurt a. M.: Vervuert 1989, pp. 85–98.

[38] Cf. on this Ette, Ottmar: Cirilo Villaverde: 'Cecilia Valdés o La Loma del Angel'. In: Roloff, Volker/Wentzlaff-Eggebert, Harald (eds.): *Der hispanoamerikanische Roman.* Vol. 1: *From the beginnings to Carpentier.* Darmstadt: Wissenschaftliche Buchgesellschaft 1992, pp. 30–43.

[39] Moreau de Saint-Méry, Louis Elie: *Description topographique, physique, civile, politique et historique de la partie française de l'île de Saint-Domingue.* 2 vols. Philadelphie 1797, p. 75. Kleist may have played with the title of this important work in his titling. I thank Gesine Müller for pointing out this connection.

[40] Cf. on this Loster-Schneider, Gudrun: *Toni, Babekan and Homi Bhabha? On Problems of Cultural and Aesthetic Hybridization in Heinrich von Kleist's 'The Betrothal in St. Domingo',*

contemporarily in other parts of the Caribbean. Kleist, however, concentrated on the yellow in the color spectrum.

How established the concept of the color yellow was as a designation for a situation among the *mulâtres* in the francophone region of the Caribbean may also be demonstrated by the following evidence:

> The basis of this racial nomenclature was to underline as 'scientifically' as possible that the mark of black blood remained forever ineffaceable. Thus, the freemen are named by the whites sometimes 'mixed-bloods', sometimes 'people of color', sometimes 'mulattoes' and, later, they will be called 'yellow' by the blacks of Saint-Domingue. As for the appellations of "métis" or "métive", they are reserved for the very rare descendants of the Indians. In the Windward Islands, one does not bother with so many subtleties: the free are generally 'mulattoes'.[41]

Incidentally, the terminological boundary between mestizo and mulatto could not have been drawn more clearly here. The fact that this biologistic distinction, primarily related to skin color, could be culturalized in the nineteenth century and transferred to Haitian literature, which was in the process of being created, is shown by the category of *littérature jaune*, to which a study published in two parts in the renowned Parisian *Revue des deux Mondes* could be dedicated under this title.[42] In this insightful article, Gustave d'Alaux attempted to give Haitian literature the right to exist and, for the first time, to do justice to it from a literary-historical perspective, although he placed 'yellow literature' under the sign of plagiarism and imitation of foreign, European models of French provenance. At the same time, however, he made the existence of this *littérature jaune*, which could be described as Haitian in the true sense of the word, a matter of record in Europe once and for all.

There can be no doubt that Heinrich von Kleist, even beyond Moreau de Saint-Méry, was well acquainted with parts of this usage of the term, which spread in the French-speaking world in the eighteenth century at the latest. And so, in his story *The Betrothal in St. Domingo*, the yellow color of the face also refers not only to a mulatto origin, but also implies cultural values and dispositions which—as the further course of the story shows—can be read—or at least should be read—by the male protagonist Gustav or August, as it were, from the skin and eyes, from the surface of the body of the beautiful young woman.

The color yellow, which we have already seen shining in an indirect way in the name of Congo *Hoango*, is repeatedly re-contextualized in the course of the narrative, for instance immediately after its first explicit appearance through the reference to the advance of "General Dessalines with 30,000 Negroes" (V 10) on Port au Prince, where "everything that wore the white color" (V 11) had now gathered in order not to be completely wiped out.

---

pp. 229 and 238.

[41] Abénon, Lucien-René: *La Révolution aux Caraïbes. Antilles 1789*. Paris: Nathan 1989, p. 61.

[42] Cf. in detail Müller, Gesine: *Die koloniale Karibik zwischen Bipolarität und Multirelationalität. Transferprozesse in hispanophonen und frankophonen Literaturen*, especially subchapter IV.4.1.1. Die 'littérature jaune' zwischen Frankophilie und Plagiat.

This is the situation in the final phase of the Haitian Revolution, briefly and succinctly sketched in a few strokes in *The Betrothal in St. Domingo*, when the actual events reported in Kleist's narrative begin with a stranger's nocturnal knock on "the back door" of the former mansion of Guillaume von Villeneuve (V 11). To Babekan's question at the beginning of this 'night play', "who is there?" (V 11), the stranger, invoking "Mary and all the saints" (V 11)—that is, *Toussaint*[43]—replies with a counter-question, "are you a Negress?" (V 12) With a clever riddle of her answer, the mulatto opens the gate: and with this *overture*, which as it were inscribes the name of the famous prisoner from the fortress of Joux in the text, the plot core of Kleist's revolutionary play begins.

The masterly way in which Heinrich von Kleist was able to translate macro-historical developments into micro-historical and, above all, micronarrative movements is exemplified by *The Betrothal in St. Domingo*. The fact that the alien white man reaches out with his hand to Babekan in the pitch dark of the black night in order to grasp bodily and haptically has to do with the question that haunts him until his own death, his suicide, which divides the world into white and black and wants to see everything identified only with these colors. The divorce between white and black characterizes-as we saw-the narrator's historical situating of the event in the context of Dessaline's march on Port-au-Prince. This black and white dichotomy, however, as will soon become clear, is neither sufficient to adequately understand one's own situation, nor to ensure one's own survival of the stranger who comes from a completely different cultural sphere.

No wonder, then, when Babekan's self-designation as a mulatto and the designation of her daughter—now not in the narrator's voice, but by the mulatto herself—as a "mestizo" (V 12) confuses the young Swiss Gustav and plunges him into great uncertainty. But when he tries to escape—enlightened by Nanky, an illegitimate son of Congo Hoango, about the new ownership of the mansion—Toni, in her capacity as decoy, holds him back by cleverly illuminating her face in white with a lantern. He will not be able to answer the question with which he addresses "her young lovely figure" (V 15) until her death, as well as his own: "Who art thou?" (V 15).

Toni succeeds in "dragging the white man away with her" (V 15) and into the interior of the house with her by the hand. The young woman plays the role of her life and death assigned to her by Congo Hoango and Babekan—her role as a decoy to keep white people in the house, where Congo Hoango can murder them in peace of mind—with great perfection and routine. And yet, in her tearing away, there is already a tearing away, for mixed in with her enticements is the so ambivalent talk of a black "ravager" who indeed bears the name Congo Hoango (V 15), but who is not currently in the house, as he is supplying Dessalines' army with lead and powder. The coincidence of Toni and Gustav's chance meeting begins to open onto the possibilities and risks of future conviviality.

The stranger, who identifies himself as Gustav von der Ried and as an "officer of French power, although […] not a Frenchman" (V 17), quickly wants to continue

[43] The invocation is repeated in Gustavus' formula, "God and all the saints," found just a few pages later (V 17).

his way from Fort Dauphin, where all the whites have been murdered, "through the middle of a Moorish country in revolt" (according to Babekan) (V 17) to Port-au-Prince. He does not intend to do this alone, but together with a party of twelve, including his grandfather Strömli, his wife and five children. The Swiss hopeful entrusts all this to the aging mulatto Babekan, because "a ray of mine" gleams at him "from the color" of her face (V. 18).

The wild landscape of an island, an isle, partly used by plantations and partly undeveloped, is repeatedly evoked, whereby the insularity appears here as isolation, as a prison from which one can only save oneself from death by escaping. Within this self-contained *island world*, the house from which Gustav hopes to save himself and his family forms an island of refuge within a country hostile to the whites, whereby this house appears from the beginning from a vectorial point of view. It forms the junction within an island shaken by revolutionary movements, is itself a testimony to that overthrow of conditions which culminated in the murder of the slaveholder and his family, but at the same time lies on the great highway which alone still allows access to the very last island of refuge for the whites: the port city of Port-au-Prince, secured by the rest of the French troops. The space of the house is thus not static, but a highly vectorized space of movement, in whose insularity all historical, political, and interpersonal movements intersect.

Surprisingly quickly, this house also becomes a vector field for the love that develops between Gustav and Toni in moments of supreme threat. The danger of death eroticizes life. With the "lovely young mestizo who opened the house to me" (V 20), Gustav is connected from the beginning by a desire that turns into lust after only a few hours in the house. If early on Gustav wrapped his "arm gently around her body" at the banquet Toni prepared, "pressing her briskly to his breast" (V 26), she, in turn, is quick to reciprocate this desire. And yet, just a moment ago, she had wanted to draw him to her doom as a decoy, a *femme fatale*, as it were. Toni's in-between position, her life in a multifaceted in-between, are evident.

Interposed between the night of the encounter and the night of love, however, is the story told by Gustav to the beautiful Toni of that "girl" from the "tribe of Negroes" who, at the time of the uprising, had fallen ill "with the yellow fever" that had "broken out to double the misery in the city" (V 31). From a narratological point of view, the story of this black slave in turn forms a doubling, insofar as, in relation to the main narrative strand, the color yellow already establishes a direct link between this slave, mistreated by her former master, and the beautiful Toni with her skin colour playing into the yellowish. In revenge, the slave lures her former master into the house, seemingly surrenders to his caresses, but then, full of rage, hurls at him in bed: "a pestilent woman, bearing death in her breast, hast thou kissed: go and give the yellow fever to all those like thee!" (V 32 f).

The figural[44] composition of this inserted micro-narrative as a doubling structure is emphasized not only by the element of the house or the color yellow, but also by

[44] On the concept of *figura*, see Auerbach, Erich: Figura. In Erich Auerbach: *Gesammelte Aufsätze zur romanischen Philologie.* Ed. by Fritz Schalk and Gustav Konrad. Bern/Munich: A. Francke 1967, pp. 55–93.

the chest as the place of death—Gustav will later shoot Toni in the chest, into the loving heart, not by chance. Moreover, yellow fever, which is the leading epidemic of the second phase of accelerated globalization (just as the disease AIDS stands paradigmatically for the fourth phase), continues those imaginings that were tied to the leading epidemic of the first phase of accelerated globalization. For at the turn of the sixteenth century, as many later accounts put it, Europeans had received syphilis from Native American women in revenge for the genocide they had perpetrated against the indigenous population, a disease that quickly spread—as Cornelius de Pauw, for example, pointed out in his highly influential *Recherches philosophiques sur les Américains*—to the Old World. The connection between the atrocities committed by whites, the revenge of nature and the lustful union with doubly colonized dependent women proves to be an obvious line of interpretation when compared with this writing of the Amsterdam philosopher, first published in Berlin in 1768:

> After the swift massacre of a few million Indians, the atrocious conqueror felt himself afflicted by an epidemic disease, which, by attacking both the principles of life and the sources of generation, soon became the most horrifying scourge of the habitable world. Man, already burdened with the burden of his existence, found, to make matters worse, the seeds of death in the arms of pleasure and in the bosom of enjoyment: he thought himself lost without a resource: he believed that irritated nature had sworn his ruin.[45]

In the arms of lust, at the breasts of voluptuousness lurks the danger of deadly infection: the semanticization of the woman's[46] body has a long colonial prehistory,[47] to which Heinrich von Kleist also draws attention. The nameless black slave, whose story is narratively inserted by Gustav, configures in a first doubling structure the male fears that, for the Swiss, are associated with the female lure into the house and with the mulatto woman who is so efficient as a decoy.

It is no coincidence, however, that this is coupled in the cleverly constructed plot with the danger of infection by yellow fever, since Kleist had referred to this rampant disease of the *fièvre jaune* in his article 'Kurze Geschichte des gelben Fiebers in Europa' (Short History of Yellow Fever in Europe), published in the *Berliner Abendblätter* in January 1811, with the term an "occidental plague".[48] His short

[45] Pauw, Cornelius de: *Recherches philosophiques sur les Américains, ou Mémoires intéressants pour servir à l'Histoire de l'Espèce humaine.* 2 vols. Berlin: Chez Georges Jacques Decker, Imp. du Roi 1768–1769, here vol. I, p. a3v. Cf. in this context also Ette, Ottmar: 'Not just brought about by chance': reflections on globalisation in Cornelius de Pauw and Alexander von Humboldt. In: *Studies in Travel Writing* (Nottingham) XV, 1 (February 2011), pp. 3–25.

[46] Cf. Weigel, Sigrid: Der Körper als Kreuzpunkt von Liebesgeschichte und Rassendiskurs in Heinrich von Kleist's Erzählung 'Die Verlobung in St. Domingo'. In: *Kleist Jahrbuch* (Stuttgart/Weimar) (1991), pp. 202–217.

[47] Cf. on this Hölz, Karl: *Das Fremde, das Eigene, das Andere. Die Inszenierung kultureller und geschlechtlicher Identität in Lateinamerika.* Berlin: Erich Schmidt Verlag 1998.

[48] Cf. Kleist, Heinrich von: Kurze Geschichte des gelben Fiebers in Europa. In: *Berliner Abendblätter* (Berlin) 19 and 20 (January 23 and January 24, 1811), pp. 73–75 and 77–79, here p. 78. Cf. on this Loster-Schneider, Gudrun: Toni, Babekan und Homi Bhabha? On Problems of Cultural and Aesthetic Hybridization in Heinrich von Kleist's 'The Betrothal in St. Domingo', p. 239.

study in medical history, which appeared a few weeks before the first publication of the narrative examined here, had shed particular light on the great yellow fever epidemics of 1793 and 1804, both of which not coincidentally fall within the period of the Haitian Revolution and thus within the diegesis of *The Betrothal in St. Domingo.*

Kleist's reflections document his awareness of the extent to which yellow fever had become a disease and plague, whereby the writer was interested not only in the symptomatology of this 'occidental plague' but also in its worldwide spread. Kleist's narrative repeatedly plays with phenomena of world-historical significance, but at the same time plays them out micro-historically at the level of his narrative.

The outbreak of yellow fever among the French troops under General Leclerc—who himself succumbed to this disease in the course of the war as early as November 1, 1802—may well be regarded on the level of historical background as an important, perhaps even war-deciding event. A transfer to the plot level of Kleist's narrative also occurs insofar as it is the "officer" Gustav who asks Toni "whether *she* would be capable of such a deed" (V 33). Even if Toni clearly denies the question: when Gustav enters the room assigned to him, which had probably belonged to Guillaume von Villeneuve before, he clearly notices how "a feeling of restlessness settled around his heart like a Geyer" (V 35).

But the view of the "dark curls" and "the young breasts" of the girl (V 35) who prepares the washing of the stranger's feet distracts Gustav from all premonitions of death and cheers him up again. In mortal danger, he rejoices in the spectacle of life. And yet, as far as Toni's beguiling beauty is concerned, there is a qualification, for "he might have sworn, except for the color, which was offensive to him, that he had never seen anything more beautiful" (V 36). Did not Toni's yellow complexion here blend dazzlingly with the yellow fever of her black *figura*?

A short time later, Gustav notices a "distant resemblance" (V 36) which he does not know how to flesh out at first, but which introduces a second doubling structure in which the *figure* of Toni is placed, as it were, between the *figura* of the black slave with yellow fever and the *figura* of a white girl. A fifth and final triangle emerges, personalizing the triangle of black, white and mulatto.

For soon the stranger will turn out to be none other than Gustav's bride, who died under tragic circumstances. As the stranger pulls the fifteen-year-old girl onto his lap and clasps her with his hands, a familiar, almost intimate conversation ensues, during which Toni finally nestles against the strange man's breast "with an exceedingly charming blush that flamed up over her burnt face" (V 38). After a first shy kiss on her forehead (V 39), he recognizes in her, as in anagnorisis, the image of that young woman who had once sacrificed herself for him as his dear bride. Once again Toni is connoted with death, but also with love and lust.

The interwoven micro-narrative about Mariane Congreve, Gustav's amorous bride from Strasbourg, who was once innocently guillotined in his place, places the image of Toni *between* those of the black slave and the white fiancée, so that the mulatto comes to the fore as an object of male desire oscillating between the two poles. Like the nameless slave, the image of Mariane Congreve appears as that of a doomed woman, a male projection of female images (since always narrated by

Gustav)[49] that can only be conveyed through her beautiful corpse, which, thanks to the aforementioned figural superimposition, is necessarily transferred to Toni as well.

If Gustav had been imprudently critical of the revolutionary tribunal of the recently established French Revolution, Mariane was executed in place of her fugitive bridegroom—not without being able to cast a last, indelibly imprinted glance from the scaffold at the horrified Gustav, who was no longer able to save her (V 42). Mariane sacrificed herself by publicly denying that she even knew Gustav, who had become guiltless. And will Toni not deny herself in order to save herself and Gustav from betrayal after all? In Kleist's game of deception and dissimulation, however, death usually has the last word.

Only now, after the doubled doubling, can the narrator tell of the night of love. By mixing her tears for Mariane Congreve with Gustav's (V 43), Toni allows the bodily fluids of both to flow together in an all-too-brief night of love, and at the same time assumes the figural legacy of the executed Strasbourg bride. Will Gustav not bring death to her as well? It is thereby first of all the little death. The mulatto's intermediate position, laid out early on, is fulfilled in a union that, in the middle of the narrative, need only be narratively executed by silence: "What further took place we need not report, because everyone who comes to this place read't of his own accord." (V 43).

Afterwards, Toni's tears can hardly be stilled. The far-reaching consequences of that "deed which he committed" (V 43) remain unclear to the Swiss at first; and yet that gesture with which he presents Toni with the golden cross of Mariane Congreve as a gift is not only the handing over of a piece of jewelry, but far more the conclusion of a contract that Gustav had originally made with his white bride.

It is, as will soon become clear, a contract with the barely fifteen-year-old mulatto, which remains hidden, with which a future coexistence across racial barriers is agreed. It is a pact of equals, but one in which the Antillean woman, connecting America, Africa and Europe, soon proves to be not only her partner's equal, but clearly superior.[50] For it is only through her cunning that the chance opens up to counter the colonial burden with a pleasure to be enjoyed in the future.

A Vision of Conviviality

The fact that Mariane Congreve comes from Strasbourg, that 'city of streets', is probably no coincidence, because her heiress Toni also owes her chance encounter with Gustav von der Ried to her house on the great country road. Both encounters are marked by the transitory, a transit from which it could not be assumed that a continuation of the only temporary cohabitation would ever be conceivable. For in

[49] Cf. Bronfen, Elisabeth: *Nur über ihre Leiche. Tod, Weiblichkeit und Ästhetik.* Munich: dtv 1994.

[50] See also Fleming, Ray: Race and the Difference It Makes in Kleist's 'The Betrothal in St. Domingo'. In: *The German Quarterly* (Riverside, California) LXV, 3–4 (summer–fall 1992), pp. 306–317.

the midst of the military conflict of the colonial antagonists, in the midst of the trial of different gender identities and ethnic origins, how could one seriously expect that, despite all exclusions, all marginalization and massacres, there would be a chance of living together? Isn't the game that Toni and Gustav have embarked upon a lost game from the very beginning, one that must end in blood?

Maybe so. But Kleist lets them act it out and live it out in the testing room of his narrative. Only in this way does *Die Verlobung in St. Domingo* become the many-voiced experimental space of a cohabitation that is basically unlikely to have any chance of future. The "bridal gift" (V 43) makes it clear that the sexual union of white man and mulatto woman, of colonial master and colonized woman, represents an engagement that would only be able to subvert the traditional patriarchal making of the mulatto woman available for male lust through the project of future cohabitation. For from a colonialist point of view, isn't the black woman there for work, the white woman for marriage, but the mulatto woman for quick sexual gratification?

The contract agreed upon between Gustav and Toni, their 'betrothal', holds against this colonialist planter ideology Gustav's promise to ask his mother for her hand in marriage the very next morning, and immediately transitions into a vision of future conviviality, of idyllic cohabitation in a supposedly perfect world of Europe:

> He described to her what a small property, free and independent, he possessed on the banks of the Aar; a dwelling, comfortable and spacious enough to accommodate her and also her mother, if her age permitted the journey; fields, gardens, meadows and vineyards; and an old venerable father, who would receive her there gratefully and lovingly, because she had saved his son. (V 44)

This vision gives rise to an other-place, an idyllic heterotopia that bears all the traits of a place of pleasure, a *locus amoenus*. It is the idyll of a world regulated by private property and patriarchal goodness, which opposes the frenzy and murderousness of a colonial world that has come apart at the seams. Even if Gustav swears to his Toni that he only took possession of her "in the frenzy of wonderfully confused senses, a mixture of desire and fear that she instilled in him" (V 44), and not out of masculine calculation, his counter-image of a permanent marital union is nevertheless well-calculated testimony to a rationally designed world, which itself represents a small, self-contained island of bourgeois bliss that could hardly stand in sharper contrast to the situation in Saint-Domingue.

And yet, in a highly ambivalent way, a not insignificant bit of Saint-Domingue creeps into this idyll of a Swiss otherworld. For is the mulatto not honored here by Gustav's father out of gratitude for having saved his son's life by accepting her into the family? In a further doubling structure, is the sign of great gratitude not celebrated as a patriarchal benefit in the same way that Congo Hoango was once rewarded for saving his master with release, financial means, the mulatto life partner Babekan and inclusion in the circle of those who are allowed to exercise authority over slaves?

It therefore seems no exaggeration to claim that a colonial birth defect has crept into Gustav's happy vision of future conviviality, confronting Toni, a mulatto, while still in Switzerland with those colonial asymmetries that she would have preferred to leave behind entirely. The colonial germ is still in the European idyll.

Nevertheless, Gustav's Swiss idyll is the counter-image to a colonial society whose collapse had already been translated into a violent image of the body that cannot be stopped because it is fragmenting:[51] "Is it not," Babekan had said, "as if the hands of one body or the teeth of one mouth wanted to rage against each other, because the one member is not created like the other? (V 19 f) In the society of the merciless European slaveholders as in the counter-society of a Haitian revolution, which only reverses the polarity of a community shaped by racial hierarchies, mechanisms of exclusion bloodthirstily confront each other in Kleist's narrative in mirror-symmetrical exclusivity, which do not spare even one's own body in the image of the body torn and torn within itself.

Against all images of horror, Gustav von der Ried—who does not share the title of nobility with Herr von Villeneuve (but also with Kleist himself) for nothing—sets the Swiss vision of an inclusion, which in a special way undertakes the inclusion of the beautiful mulatto in the circle of the family. Thus a connection transgressing racial barriers between the white officer and the fair-skinned mulatto becomes concretely imaginable, so that at this point in the narrative, after the sexual union of Toni and Gustav, an alternative model for coexistence between different ethnic groups begins to emerge. The overcoming of racial barriers becomes imaginable, conceivable, perhaps even livable. Heinrich von Kleist explores the opportunities and limits of his experimental play.

Certainly, this model is relocated to a distant Switzerland, where the fields, gardens, meadows and vineyards sketch a landscape of theory, where, under the sign of general fertility, opposites no longer clash abruptly, but are sweetly balanced. The conviviality between different genders and ethnic groups, between different cultures and classes becomes imaginable, even if Kleist's narrative uses the example of the Haitian Revolution to demonstrate how far the road still is to the realization of such a conception of peaceful coexistence—no matter how patriarchally tinged it may be. Love is certainly an always precarious means of movement here, since it can also turn into hatred in the case of misunderstandings, which—as Kleist impressively demonstrates in his story—has a destructive and self-destructive effect. And yet: an engagement is emerging between the two lovers that promises not only happiness for the future, but lived conviviality.

The polylogically unfolding ZusammenLebensWissen in *Die Verlobung in St. Domingo*[52] sketches a profound, certainly also tragically tinged picture of conviviality on an individual as well as on a collective level. It is fascinating to see how Heinrich von Kleist does not only sharpen and historically charge the modernity of his view—like most of his German contemporaries—with a critical retrospective on

[51] Cf. on the dismembered body in Kleist the considerations in Jauß, Hans Robert: Befragung des Mythos und Behauptung der Identität in der Geschichte des 'Amphitryon'. In: Hans Robert Jauß: *Aesthetic Experience and Literary Hermeneutics*. Frankfurt a. M.: Suhrkamp 1982, pp. 534–584.

[52] On this concept cf. Ette, Ottmar: *ZusammenLebensWissen. List, Last und Lust literarischer Konvivenz im globalen Maßstab (ÜberLebenswissen III)*. Berlin: Kulturverlag Kadmos 2010.

the European double revolution. He does not limit himself to including reflections of the political revolution emanating from France and on the industrial revolution emanating from England, but offers his readers a world-historical dimension by making the Haitian Revolution, at the crossroads of a wide variety of discourses, the site of reflection for his desperate love story. Doesn't the driving force of love bring to light all those contradictory factors that are crucial for living together on a global scale, a conviviality that precisely does not want to be limited to a Swiss idyll?

The past of the tyranny of a merciless plantation economy, like the present of a violent, cruel liberation of its victims, holds both lovers firmly in its claws. Gustav makes it clear to Toni that he is not willing to "protect" the "general relationship" of whites to blacks—by which he euphemistically refers to slavery—without reservation (V 31). But misdeeds of "manifold and reprehensible maltreatment" had been committed only "by some bad members" of the sugar oligarchy (V 31), so that all this could not justify the "madness of liberty" and the revenge of blacks on whites for their system, which had existed "for many centuries" (V 31). How could the slaves, how could a Babekan who had not only been disowned by her daughter's father but also mistreated and whipped by Lord of Villeneuve (V 29), have agreed with such an assessment?

In the narrative, many different discourses become audible under the narrator's voice, so that in Bakhtin's sense one can certainly speak of a cosmos of speech diversity that reacts highly sensitively to the challenges of the second phase of accelerated globalization. In *Die Verlobung in St. Domingo*, Heinrich von Kleist has succeeded in modelling a literary testing ground out of this diversity of speech and voices, the focus of which is the question of the opportunities, but also the risks and dangers, of future coexistence on a global scale.

Heinrich von Kleist's narrative thus unfolds a broadly conceived ZusammenLebensWissen, which cannot be limited or curtailed in its diversity of meaning by any single narrative voice. The *polylogical* narrative structuring, committed to many different logics at the same time, demonstrates the perspectives that violently collide in the burning mirror of the Haitian Revolution. Not only for Humboldt, but also for Kleist, Haiti becomes a challenge, indeed a paradigm, against which the future in a global context cannot only be assessed, but also narratively tested.

However strong and destructive this clash in the second half of the narrative may be, it is unable to conceal what must be called the prospective, future-convivial dimension of Kleist's narrative.

Conviviality and Catastrophe

If at the end of the first and only night of love the beautiful mulatto appears "like a lifeless one" (V 45)—a small death that already foreshadows the great one—the new day brings up Babekan's plan to hold on to the travelling party of whites by deception and dissimulation until Congo Hoango can lead these white or Creole half-dogs, as he calls them (V 21 f), like so many before them, to their deserved fate

(V 46). The image of pleasure is displaced by the life-threatening burden, the image of heterotopic conviviality by that of topical confrontation: Everything threatens to sink into a bloody catastrophe. But with the help of what new cunning might it be possible to regain a lasting pleasure in conviviality?

Toni immediately rebels against her mother's plan to turn the "house"—as she has done too many times before—into a trap, a "murder pit" through a system of well-calculated lies (V 47). Her love makes her inventive. But Babekan points to earlier examples of how not only the French but also the Portuguese or the Dutch (and thus the representatives of other European colonial powers) were successfully hunted down (V 48 f)—quite independently of any individual guilt in each case, and without anything concrete being able to be "laid to their charge" (V 49).

Toni, however, does not want to be reminded of "these atrocities" by her mother, as she has long since rebelled against being forced to participate in these almost ritualistic murders (V 49). Thus the young woman places herself protectively in front of the Swiss, swearing that she would rather die "tenfold death" than see Gustav harmed even a hair (V 49).

Then, however, Toni changes her tactics and resorts to a new ruse, apparently accepting her mother's suggestions. Babekan, however, becomes suspicious and asks what it was that made Toni change her mind so radically in favour of the stranger in the course of a single night (V 52). The daughter, however, has firmly settled on her ruse and emphasizes that it was only at the sight of her mother that "the whole inhumanity of the species to which this stranger belonged" had once again become clear to her (V 52 f). For Babekan's body still shows all the signs of the abuse she suffered in the colonial system: She is the living figuration of the brutality and inhumanity of pan-European colonialism.

For Gustav, however, Toni's behavior soon becomes puzzling. He cannot really trust what the narrator knows: that Toni no longer sees in yesterday's stranger a mere guest, but "her fiancé and husband" (V 59). Neither understands nor trusts. Yet not only for Gustav, but also for Toni, the betrothal is now complete, indeed already includes the marriage to the young officer. For the young mulatto, their single night of love has become an engagement, even a marriage, so that, like Mariane Congreve, the first bride of the Swiss, she is ready at any time to lay down her life for her husband without hesitation. But this time, too, Gustav will not understand in time.

In this way, however, the image of future conviviality is brought from the otherworld of Switzerland into the here and now of a coexistence in the Caribbean that is profoundly threatened in its existence. And this not only from the outside, from all those protagonists and groups who have set out on a merciless hunt for white people, but also from within the relationship itself. For Toni has not let Gustav in on the ruse she has chosen to transform the mortal danger weighing heavily on both of them into shared pleasure. This asymmetry of knowledge will have fatal consequences.

Once again, the house becomes the playing field for a deadly confrontation. The daughter's "cunning" is pitted against the mother's "cunning" (V 62). Toni hopes for divine support for her own plan, as she kneels before the "image of the Holy Virgin that hung beside her bed" (V 62) and implores her Son Jesus Christ, "in a prayer of infinite fervour, for courage and steadfastness", also in order to be able to confess the crimes she has committed, "which weighed down her young bosom" (V 63). This burden weighs heavily on her, but she hopes for Gustav's forgiveness and that he "will take her with him to Europe as his faithful wife" (V 63).

Thus the prayer to an omnipresent transcendence is transformed into the hope of heterotopian conviviality, Gustav von der Ried's draft of future cohabitation projected into Switzerland has evidently triggered an effect in the mulatto woman that is probably even greater than that of sexual union. Not this, but the promise of future conviviality forms the core of an engagement that represents a prospective act and pact that seeks to overcome racial barriers between yellow and white.

Gustav himself also seems to live this pact—admittedly in a dream, in which he often whispers Toni's name "with glowing, trembling lips" without waking up from the call of the real girl (V 64). Thus, in Gustav's bedchamber, Toni cannot bring himself to "drag him down from the heavens of lovely imagination into the depths of a mean and wretched reality" (V 64)—a momentous mistake, as Gustav will understand Toni's ruse too late. In this last scene before the storm, the two lovers live in two different worlds, in dream and in reality, in heaven and in miserable depth.

Consequently, the bedchamber in the house's space of movement can no longer become the testing ground for their life together. For already not only the moonlight, but also "the night wind" penetrates through the "open windows" (V 63) into the interior of the house, an intrusion that, as a breaking in, will soon play not only with the hair on Gustav's forehead, but with his whole body as well as with Toni's. The bed never again becomes the playground of their desire, but of bondage. The bed will never again become the playground of their lust, but of bondage: from now on, alien powers rule their bodies.

Only a few lines later, Congo Hoango enters the interior of the courtyard and the house, which had just promised protection, with the auxiliary squad he commands for General Dessalines, and the "sound of men, horses and weapons" becomes audible (V 64). At the same time, it becomes clear that the mother suspects her daughter's cunning and therefore already denounces her to Congo Hoango as a "traitor" (V 65).

Moreover, Babekan's daughter is in danger of Gustav himself henceforth "thinking her a traitor" who would have acted in concert with her mother and Congo Hoango (V 67). Once again the young mulatto stands between two poles, between two fields of force and danger; and her cunning, happily entered by chance, in quickly binding Gustav hand and foot even before Congo Hoango does, will consolidate in the Swiss officer the false understanding that Toni has only tied him to the bed with her charms in order to deliver him as a white man to the blacks. Lustful conviviality remains confined to the dream: Reality seizes the lovers and sweeps them away into an inescapable death vortex.

In the former house of the slave owner, in the mansion where Congo Hoango put the bullet through his former master's head, peaceful coexistence is no longer even a dream. All these movements are vectorially stored in the fractal of the mansion now inhabited by the former slave: The readers of the narrative can call up here all those trajectories that help to determine the behavior of all those involved.

In the 'stress test' of this Caribbean reality in Saint-Domingue, the project and projection of conviviality prove to be a dream that immediately threatens to turn into a nightmare. Heinrich von Kleist's narrative aesthetically savors all the details of this flip. Not only does Congo Hoango, as oblivious to Toni's wiles as Gustav von der Ried, refer to her as "his dear girl" (V 70), thus once again emphasizing her intermediate position between black and white—and this time with a view to the emotional movements she triggers. Rather, Babekan's daughter is hit hard by those looks which Gustav, still tied up in bed, had thrown at her "full of contempt" and which had immediately, "like stabs of a knife, gone through her heart" (V 72). She rightly feels how, instead of conviviality, disaster is steadily approaching. But unlike Gustav, she is not paralyzed in her actions.

For she succeeds in warning Strömli and his community of white fugitives of the trap, of the pit of murder into which the house of Guillaume of Villeneuve has again been transformed. Toni has stepped entirely into the role of Gustav's first bride and exults "at the thought of dying in this enterprise ordered to his rescue" (v 73). For conviviality increasingly appears realizable neither in this place, the island of Hispaniola, nor in the other place, the island on the Aare, but only in a transcendence, so that the small death of sexual union can only be experienced together in the great death. The narrative draws its power, indeed its violence, from this stringently maintained narrative logic.

There is a last glimmer of hope when Toni succeeds in warning Strömli and his community of the danger and persuades them to take up arms and rush to the house to save "Cousin August" (v 76). The often-debated[53] doubling of Gustav's name into August indicates the same oscillation between two names that could already be discerned in the title fractal. The hybrid bifurcation of the island corresponds to the bifurcation of a human being who will confront Toni, but also the reader, during this phase of highest tension and danger in a different form, deeply confused in his knowledge of life.

If Gustav, despite his initial hesitation, was devoted to Toni and became her bridegroom, he will now act as a stupid August who never really sees through Toni's cunning, even if he still shares with Gustav the lexeme *gust*, which etymologically refers back to Latin and thus unites French and Spanish. This refers not least to the pleasure, the taste, the sensual and aesthetic experience that *Gustav/August* still associates with the beautiful mulatto. Yet, in the doubling of the name, it moves from the initial position written in majuscule to a final position written in minuscule, calling attention to fundamental recasting, indeed a certain reversal. The dream of conviviality soon turns into disaster. Could Gustav have averted it had he remained Gustav on the morning of his death?

The conquest and occupation of the house by Strömli's small troupe is swift and raises the confrontation between blacks and whites in the microcosm of the house from the individual to the collective level: it is the struggle of indissolubly hostile

---

[53] Cf. for example Loster-Schneider, Gudrun: *Toni, Babekan and Homi Bhabha? On Problems of Cultural and Aesthetic Hybridization in Heinrich von Kleist's 'The Betrothal in St. Domingo'*, pp. 245 f.

communities in which it is necessary to choose sides, since there are no longer any shades in between. This is why Toni, as if in an act of self-disguise, describes herself to her mother, by whom she is insulted as a "low-life and traitorous woman," as a white woman who is "engaged" to the young Swiss (V 81): "I belong to the race of those with whom you are at open war, and before God I shall know how to answer for having taken their side." (V 81 f).

The catastrophe, already figuratively foreshadowed in the figures of the black slave and the white bride Mariane, occurs when "cousin August" (V 83), as he is now repeatedly called by the protagonists, but also by the narrator himself, sees Toni again for the first time after his liberation and "changes color at this sight" (V 84). Only a few seconds later, August, described by Strömli and his sons as a "monstrous man" (and thus as a hermaphrodite[54] between monster and human), will kill the beautiful mulatto woman "gnashing with rage" with a pistol shot "right through the chest" (V 84). Too late, the foolish August learns from his traveling companions that the young mulatto had been prepared "to escape with him, to whom she had sacrificed everything, parents and property, to Port au Prince" (V 85). If August had understood her cunning, with which she wanted to free herself from her burden, from her guilt for the death of others, it would have been possible, not in the Caribbean, but elsewhere, to realize the dream of conviviality, at least in distant Europe. Paradise was almost reached: but shortly before that, the gates slammed shut. Forever.

The catastrophe is immediately preceded—as we have seen—by a double change of color. While the mulatto, or, as she is deliberately called several times in the text, mestizo, changes to white, Gustav changes his name not only to August, but also his color. The narrator does not tell his readers what color the young white officer changes to: Is it the red of anger, the yellow of oscillating 'in-between' and jealousy, or the black of death? In any case, the change is doubly fatal, for when his old name Gustav is now thundered "in his ears" (V 85) and the dying Toni defends herself against the accusation of betrayal, "Cousin Gustav" (V 86) rises again from his bed, once more puts "his arms around her body" and accepts her last words "with a lamentably torn heart: "'you should not have mistrusted me!'" (V 87). When Toni breathes out "her beautiful soul" (V. 87) and has already "fled to better stars" (V. 88), the man who has become Gustav again puts the bullet "through his brain" (V. 88).

The suicide, in whose literary staging[55] Kleist scholars have often seen the anticipation of Kleist's own suicide together with the terminally ill Henriette Vogel on 21 November 1811 at the Kleine Wannsee, allows the possibility of conviviality within reach to appear in the catastrophe itself. Toni's and Gustav's hands miss each other for the last time, but in this final movement they show how close they had come to

[54] Among others, Blamberger, Günter, draws attention to the importance of hermaphrodites: *Heinrich von Kleist. Biographie.* Frankfurt a. M.: S. Fischer 2011, p. 434.

[55] Cf. also the chapter 'Economy and Anconomy of the Victim—Kleist's Last Production: the Suicide at the Wannsee' in Blamberger, Günter: *Heinrich von Kleist. Biography*, pp. 450–468.

a union. Toni lends a beautiful body to the image of the beautiful soul, certainly.[56] But Gustav would have taken this soul, this triple-discriminated body—as a 'colored', as a woman and as an Antillan—across all racial barriers into a *new* world, which admittedly was not yet to be found in the European colonies of the New World. Yet even in failure, even in dying, the livability, indeed the living through of such conviviality is impressively confirmed aesthetically.

The final agreement between Strömli and Congo Hoango to return the children of the black man in "Sainte Lüze" (V 90)—also a hybrid, bilingual toponym, of course—taken hostage in exchange for free passage, shows how contracts based on trust are possible, even if they are only the very first steps on a collective level.

On an individual level, concerning the centrally placed love story, the failed project of future cohabitation opens up to a projection that allows a cohabitation to appear beyond earthly life, since the bodies of Toni and Gustav are buried in a common grave at the Möwenweiher—regardless of their different origins—after the rings on their hands have been exchanged. The graves of the two lovers, however, are transformed into "the dwellings of eternal peace" (V 90), whereby the place of this peaceful coexistence in difference is now an other-place, a heterotopia, so that conviviality is shifted into transcendence and translated into the heavenly.

The flight to Port-au-Prince, the rescue (after the conquest of the city by Dessalines) on the ships of the (logically) English fleet, the crossing to Europe and the return home to Switzerland end, as if in fast motion, a narrative sophisticated in its narrative procedures, which does not end in the Swiss idyll of Gustav, but rather of Mr. Strömli's family. The narrator's last words are about the monument that the valiant Strömli had erected to "Gustav, his cousin, and his cousin's fiancée, the faithful Toni", which could be seen on the estate in Switzerland "as late as 1807".

The fact that the monument could *still* be found there in 1807 makes little sense in view of the few years that have passed since the reported events; what is more decisive is that this date refers to a biographeme already mentioned, to Heinrich von Kleist's direct acquaintance as a prisoner with that Joux fortress in the Jura whose dungeon had become a torture chamber and a cold grave for Toussaint-L'Ouverture. Thus, into the monument to Toni and Gustav is inserted a no less literary monument to one of the great figures of the Haitian Revolution, the anticipation, as it were, of a memorial plaque to the great Haitian, such as we find today in front of that fortress on the Swiss border. As recently as 1807, no one would have dared to place it there.

Heinrich von Kleist's story *The Betrothal in St. Domingo*, which continues to fascinate us today, unfolds from its title fractal a *many-voiced* and even more *many-logical* knowledge of life that, thanks to a refined narrative art, confronts us as a knowledge of experience that presents the knowledge of survival as well as the knowledge of living together in a period that, in the second phase of accelerated globalization, was characterized by the collapse of an outmoded coexistence based on slavery and thus on extreme inequality. In the collapse of a no longer sustainable racist conviviality, the narrative unfolds the diegesis of a New World in which,

[56] Cf. Weigel, Sigrid: *Der Körper als Kreuzpunkt von Liebesgeschichte und Rassendiskurs in Heinrich von Kleist's Erzählung 'Die Verlobung in St. Domingo'*, p. 212.

alongside the signs of destruction, the first signs of a future peaceful coexistence emerge.

Had not Simón Bolívar, who himself had to seek refuge in Haiti at the beginning of 1816 on his flight in the fight for freedom, claimed in his famous *Carta de* Jamaica of 1815 from the Caribbean neighbour island *Jamaica* that the intermediary life between worlds was decisive for the New World? Thus the *libertador*, who corresponded with Humboldt, addressing the inhabitants of his America—who represented a "small humanity" in their own right—emphasized that "we are neither Indians nor Europeans, but a middle species between the rightful owners of the land and the Spanish usurpers: that consequently we are Americans by birth and our rights are those of Europe."[57]

In the characters of his story, but especially in Toni, the mulatto woman described as a mestizo, Heinrich von Kleist has drawn that oscillation between different worlds that has been so characteristic of the Caribbean, that zone of most condensed globalization, since the beginning of European colonization. From a European perspective, this oscillation, this constant change of 'color', has always caused suspicion. Didn't Gustav fail because he wanted to assign the beautiful mulatto to either the blacks *or* the whites, but not to an inconclusive movement in the space between?

In Toni's last words addressed to "her" Gustav, in that "'you shouldn't have mistrusted me!'" (V 87), certainly no simplistic moral is drawn from the story, but that lexeme is emphasized which is echoed once again in the narrator's final words about "faithful Toni" (V 91). Certainly Toni has sided with the whites, certainly she has chosen the path to Europe where she was conceived and born (V 20), and certainly the entire narrative seems to come to rest in the idyllic "region of the Rigi" (V 91). But the different, contradictory and often antagonistic contemporary discourses that Heinrich von Kleist was able to weave into his narrative make it a testing ground for a possible conviviality, in whose failure a knowledge of future coexistence appears.

If all literature is writing and thus life after paradise,[58] then Kleist's reflections from his famous essay 'On the Puppet Theatre' could very well be related to that narrative which diegetically, with the Haitian Revolution, approached not only the contemporary paradigm and thus the question of coexistence, but also that region where Christopher Columbus had once assumed the earthly paradise. Heinrich von Kleist's writing knows that after the expulsion of Adam and Eve from paradise and thus after a first application of physical violence in the human-historical fiction of *Genesis*, the return to Eden is forbidden and seems impracticable for all time. Literature, however, in the sense of Kleist, has not yet abandoned all hope: "But Paradise is barred and the cherub behind us; we must make the journey around the world and see if it is perhaps open again from behind somewhere."[59]

---

[57] Bolí var., Simón: *Carta de Jamaica. The Jamaica Letter. Lettre à un Habitant de la Jamaïque*. Caracas: Ediciones del Ministerio de Educación 1965, p. 69.

[58] Cf. on this, among others, the opening chapter in Ette, Ottmar: *Konvivenz. Literatur und Leben nach dem Paradies*.

[59] Kleist, Heinrich von: On the puppet theatre. In: Heinrich von Kleist: *Sämtliche Werke und Briefe in vier Volumes*, Vol. III, p. 559.

Rahel Levin Varnhagen or the Mobile of Prussia

'But Everything Is Over!'

> At my tea table … only I sit with dictionaries; tea is no longer made at my place at all; so everything is different! I've never been so alone. Absolutely. Never so absolutely and definitely ennuyirt. For only spiritual, kind, hope-giving things can still hold out one so mortified, one so killed. But all is past. In the winter, and still in the summer, I knew some Frenchmen: with them I talked back and forth, and we denied what foreign-mannered, literature-loving and practicing people who are not of one country can deny and deny. They are all gone. My German friends, how long ago; how dead, how scattered![1]

A woman sits in front of the ruins of her existence, of her life. At a tea table in an apartment where no more tea is being made, because no one is there anymore, because there is no more living together: no living together and no conviviality[2] in what used to be a constantly expanding circle. All the woman's friends, all her guests, all her soulmates are gone, disappeared, blown away by the wind, dispersed. Absolute loneliness.

Where once conversation, cultivated conversation, convivial exchange were predominant, only dictionaries remain: books in which are preserved and stored those words that once made the rounds in lively interplay and illuminated an entire society. Conviviality, the joyful living together in a circle filled with sociability, has been replaced by absence, a loneliness that in its absoluteness weighs on the no longer quite young woman as never before. Conviviality and living together have given way to a yawning emptiness, a loneliness filled with grief: 'Everything is over!'

At least that is how it wants to appear. The lonely woman at the orphaned tea table is Rahel Levin, who sits in the deepest consternation, in absolute solitude in a room that had been filled with movements of both an inner and outer kind and is now forever deserted—and will remain so. Until recently, she had been the idolized

[1] Varnhagen, Rahel: Letter to Gustav von Brinckmann (1808). In: Varnhagen, Rahel: *Gesammelte Werke.* Vol. I Munich: Matthes & Seitz 1983, p. 328.

[2] On this term cf. Ette, Ottmar: *Konvivenz. Literatur und Leben nach dem Paradies.* Berlin: Kulturverlag Kadmos 2012.

O. Ette, *Mobile Prussia*, https://doi.org/10.1007/978-3-476-05839-3_6

center of a society that looked cheerfully and confidently to the future. But even the few French people with whom she was still in contact have gone, leaving her alone.

It is the (literary) rearguard of those French who marched into Berlin with Napoleon in 1806 and brought Prussian society, still very much in the post-Friderician era under the sign of tolerance and openness, to an abrupt end. Prussia itself seemed to be at an end. The time of Frederick the Great was now finally over, the spirit of Sanssouci gone: the Prussian rococo with its cheerfulness and its garlands of plants had disappeared like its soldiers and the Tall Guys. What was actually left of Prussia?

And the French? Now they, too, have moved on, eastward, no doubt, towards their own demise, the demise of the *Grande Armée*, victorious in the light of the burning villages and small towns of Russia. But the Prussian friends and guests have also gone, have either fallen like the Hohenzollern Prince Louis Ferdinand at Saalfeld in the fight against the French, or have come between the fronts of Prussia and the French like Adelbert von Chamisso; have preferred to remain in Paris when Napoleon marched into Berlin, like Alexander von Humboldt, who was busy with his American travel work, or are soon entrusted with the reorganization of the Prussian state like his older brother Wilhelm; or they are thinking of resigning from the Prussian civil service and leaving Königsberg again, like Heinrich von Kleist, who soon became a prisoner of war in France, or Friedrich von Gentz, who was devoted to his deeply anti-French attitude and aligned himself with Austria. They all belonged, as a matter of course, to the circle that had gathered around Rahel Levin. And they were gone.

The traces of many, which are gathered in the present volume, point in all directions: Prussian society had to find and reinvent itself before a new beginning, which after all at its core meant not regeneration but restoration, could be tackled. In the Napoleonic whirlwind, all the figures rearranged themselves before the richly embroidered chairs of the Congress of Vienna finally formed the new order of monarchies. Wilhelm von Humboldt also had his own chair here:[3] Prussia was back *on the map*.

But with the entry of the French army into the Prussian capital, the old society of Prussia and the gay society of Rahel Levin had perished together and irretrievably. It is true enough: 'But all is over!' Nothing could bring back and revive these societies. Everything that Rahel Levin's so-called salon had stood for between 1790 and 1806, everything that it had not yet made socially acceptable, but at least socially acceptable—that is, the deliberate overcoming of barriers that were as national as they were religious, as 'racial' as they were class or social—was gone once and for all. Absolutely.

The atmosphere and logo-sphere had changed. The rising hatred of the French was accompanied by a never-quenched anti-Semitism in Prussia and, at the same time, by an incandescent nationalism, a combinatorics that was to determine the fate

[3] This chair by Wilhelm von Humboldt is one of the objects in the new exhibition dedicated to the Humboldt brothers, curated by Bénédicte Savoy and David Blankenstein at the German Historical Museum to mark the 250th anniversary of Alexander von Humboldt's birth.

of Prussia and Germany for a century and a half. Almost overnight, the world of her very own circle, her attic society, had become for Rahel Levin the 'world of yesterday', a suddenly unattainable dream[4] of a community that was to be realized in the tea party around Rahel for over a decade in an almost model-like manner. What remained of this dream? A woman sits in 1808 before the ruins of her existence. But also her life?

Hannah Arendt would certainly have answered this latter question in the affirmative. In her book *Rahel Varnhagen. Lebensgeschichte einer deutschen Jüdin aus der Romantik*,[5] which she had begun in 1929 at the same time as completing her dissertation with Karl Jaspers on the concept of love in Augustine, and which, with the exception of two missing chapters, she had completed by the time she fled Nazi Germany in 1933, writing down the final chapters in French exile before publishing first an English-language[6] version in 1957 and then a German-language version in 1959, she placed her Rahel entirely in the light of Judaism from the very beginning. What seems self-evident to us today was quite innovative a century after Rachel's death (Fig. 1).

Right from the *incipit* of her influential volume, she made it clear that Rachel was part of a story that went far beyond her—and that she also understood her own life story in just this way:

> What a story!—A refugee from Egypt and Palestine, I am here and find help, love and care from you! ... With sublime delight I think of this my origin and this whole connection of fate, by which the oldest memories of the human race are connected with the newest situation of things, the farthest distances of time and space. What for so long a time of my life was for me the greatest shame, the bitterest sorrow and misfortune, to be born a Jewess, I would not want to miss that now at any price'. Thus Varnhagen reports from Rachel's deathbed. It took her sixty-three years to learn what began 1700 years before her birth, took a decisive turn at the time of her life and came to a temporary end a hundred years after her death—she died on March 7, 1833.
>
> It may be difficult to know one's own history when one is born in Berlin in 1771 and this history begins 1700 years earlier in Jerusalem. If one does not know it, and if one is not exactly a rascal who always acknowledges the given, lies about the adverse and forgets the good, then it takes revenge and in all its grandeur becomes a personal fate, which is no pleasure for the person concerned.[7]

In her influential biography of Rahel, which has indeed shaped the reception history of the person she portrayed to this day, Hannah Arendt places Rahel in the larger context of a history of Judaism, to which the Königsberg-born philosopher herself

[4] Cf. on the importance of the dream Hahn, Barbara: *'Im Schlaf bin ich wacher.' Die Träume der Rahel Levin Varnhagen.* Frankfurt a. M.: Luchterhand 1990.

[5] Arendt, Hannah: *Rahel Varnhagen. Lebensgeschichte einer deutschen Jüdin aus der Romantik.* Munich/Zurich: Piper 1981.

[6] I have before me: Arendt, Hannah: *Rahel Varnhagen. The Life of a Jewess.* First complete edition. Edited by Liliane Weissberg. Translated by Richard and Clara Winston. Baltimore/London: The Johns Hopkins University Press 1997.

[7] Arendt, Hannah: *Rahel Varnhagen. Lebensgeschichte einer deutschen Jüdin aus der Romantik*, p. 15.

Fig. 1 Rahel Levin [Varnhagen] "Rahel Lievin Berlin 1807", 1807, pencil on paper, 7 × 9 cm, Paris, Bibliothèque du Conseil d'Etat, Album Houdetot, no. 124

also belonged. Just as she provisionally concluded her study, originally intended as a habilitation thesis, a hundred years after Rahel's death and provided it with a multitude of autobiographical windows that artfully turned her biography into an autobiography doubled as a survivor's writing,[8] so she understood Rahel as well as herself on the background foil of a Jewish history and a Jewish fate, of which the last chapter heading reads: "One cannot get out of Judaism."[9]

This heading refers to Rahel Varnhagen, but is at the same time also directed at Hannah Arendt and her discursive actions[10] themselves. Behind the life of one Jewess, the life of the other Jewess can always be discerned: Both share a fate that gives their lives meaning and makes sense: a doubled life in each case, inscribed in the long history of Judaism.

It is striking that Hannah Arendt[11] is not concerned with the *condition féminine*, not with femininity and even more with her specific situation and marginalization as

[8] Cf. the sixth chapter devoted to Hannah Arendt's book on Rahel Varnhagen, 'Emigration. Life over Life: Survival Knowledge from Doubling' in Ette, Ottmar: *ÜberLebenswissen. Die Aufgabe der Philologie*. Berlin: Kulturverlag Kadmos 2004, pp. 171–188; and Sánchez, Cecilia: Sobrevivir: transiciones del cuerpo mortal e inmortal en Hannah Arendt. In: Vatter, Miguel/Nitschack, Horst (eds.): *Hannah Arendt: sobrevivir al totalitarismo*. Santiago de Chile: LOM Ediciones 2008, pp. 77–88.

[9] Arendt, Hannah: *Rahel Varnhagen*, p. 201.

[10] See Lobo, María Fátima: *Hannah Arendt y la pregunta por la relación entre el pensamiento y la acción*. Buenos Aires: Editorial Biblos 2013.

[11] Cf. generally Brunkhorst, Hauke: *Hannah Arendt*. Munich: Beck 1999.

a woman,[12] but focuses solely on the *condition juive*. From this perspective, the abrupt end of Rahel's Salon is also synonymous with the fall from a high point that the 'German Jewess from Romanticism', in Hannah Arendt's sense, could never again reach in her life. From this perspective, everything that had to do with Karl August Varnhagen von Ense, with his self-sacrificing admiration, with his marriage to Rahel, and even with his care of her estate and her writings, had to stand in the sign of a descent, an incomprehension, even a failure to comprehend, of which she accused Varnhagen with lasting consequences. For Varnhagen, who was only subsequently nobilized after the wedding, was stylized as something like the stupid fool in the Jewish life story of Rachel.

But did this have to be so? Hadn't Karl August Varnhagen von Ense appeared with his admiration and love on the horizon of her life and writing in the very year 1808 in which we saw Rahel Levin before the ruins of her existence, had he offered her such a long marriage after her conversion to Christianity and finally also offered her the opportunity to open her second salon in Berlin and thus to continue the threads of her life, in short: to become the Rahel Varnhagen that Hannah Arendt herself used to call her? Is this not how the complex figure of Rahel Levin Varnhagen emerges, whose dazzling contradictions can only develop to their fullest extent?

No: in 1808, not everything was over by a long shot: on the contrary, conditions developed that allowed Rahel to become what opened up to the 'memory of her friends' and made her a great figure in the history of German-language literature and ideas.

In the Salon of Rahel Levin

Rahel Levin was born in Berlin in 1771 into a wealthy Jewish merchant and banker family as the eldest child and grew up in sheltered circumstances. Her father, Markus Levin, was a jeweler and skilled in money transactions. As patriarch, he guided the fortunes of his wife Chaie and his family strictly, but always within a framework in which conviviality and living together were highly valued. The children were always accustomed to a house full of guests and many conversations: sociability was a high good, which the young Rahel also always tried to preserve.

After the sudden death of her father in 1790, the takeover of the business by the male descendants and the rapid settlement of all financial matters, a salon was established in the same year in Berlin's Jägerstrasse, as the centre of which emerged the young Rahel Levin, who, although not very pretty according to contemporaries, was immensely perceptive and witty. It was one of those Jewish salons that had also

[12] Cf. for a critique of this 'blank space' of Arendt's in particular Maslin, Kimberly: Rahel Varnhagen: Arendt's Experiential Ontology. In: *New German Critique* (Durham) XL, 119 (Summer pp. 77–96); and from another perspective Benhabib, Seyla: The Pariah and Her Shadow: Hannah Arendt's Biography of Rahel Varnhagen. In: Honig, Bonnie (ed.): *Feminist Interpretations of Hannah Arendt*. University Park: The Pennsylvania State University Press 1995, pp. 83–104; and (this): The Pariah and Her Shadow: Hannah Arendt's Biography of Rahel Varnhagen. In: *Political Theory* XXIII, 1 (February 1995), pp. 5–24.

formed around other women of Jewish descent such as Henriette Herz[13]or Dorothea Schlegel and which stood for that intellectual openness that was so characteristic of Berlin at the end of the eighteenth century. A free spirit prevailed in the salons, which had also taken hold of traditional gender roles. For Jewish women in particular, this was an opportunity to experiment with other lifestyles and life practices and to become visible.

The young Rahel Levin took advantage of this generally tolerant atmosphere in the Prussian capital and formed a rapidly growing circle around herself, which today is called a 'salon' in reference to Parisian structures and customs. Even if the emerging Jewish salons in Berlin differed greatly from their Parisian models, this long naturalized term will be used here for the form of sociability that centered on a woman as salonière.

Thus was born the first salon of Rahel Levin, who could not compete with the beauty of a Henriette Herz, but whose quick perception, quick wit and openness of thought inspired all those who soon learned to appreciate her diverse impulses and suggestions. And the venue was also carefully chosen. For it was not in the social rooms that her father had once used for his convivial rounds, but in the attic room that[14] people now met in deliberately simple surroundings to talk about everything that concerned the capital and its inhabitants, but also Europe and the world. The small society developed into a model of Prussia *en miniature*—at least of a Prussia that had access to education and understood education not as a possession but as a constant, uninterrupted endeavor.

Little Rahel knew how to attract artists as well as writers, intellectuals as well as politicians, scientists as well as philosophers, and how to create an atmosphere of conversation over tea and sandwiches in her 'attic room' that guaranteed—and at the same time demanded—a maximum of exchange and freedom of thought for all her guests. She brought together everything that had the right to be heard in Berlin.

The Schlegel brothers, the Humboldt brothers, Jean Paul, Ludwig and Friedrich Tieck, Prince Louis Ferdinand of Prussia and his beautiful mistress Pauline Wiesel, Friedrich de la Motte Fouqué and Friedrich von Gentz, Johann Gottlieb Fichte, Adelbert von Chamisso and such a dense circle of aristocrats and bourgeois, scientists and artists met for tea—to name just a few of the spirits of the time—that a social experiment could take place here: To live Prussia as a model, Prussia as a configuration of ideas, thoughts and philosophical themes.

It was precisely the social "spirit of optimism"[15] between 1795 and 1805 that formed the decisive background for a development in which Rahel Levin's salon

[13] Cf. Arendt, Hannah: *Rahel Varnhagen*, p. 39 f.

[14] On the origins of Rahel Levin's first salon, see Danzer, Gerhard: Die Dachstubenwahrheiten der Rahel Varnhagen. In: Danzer, Gerhard (ed.): *Europa, deine Frauen. Beiträge zu einer weiblichen Kulturgeschichte.* Berlin and New York: Springer 2015, pp. 57–70.

[15] Danzer, Gerhard: Rahel Varnhagen—eine Frau entdeckt sich über die Kultur. In: Pilz, Elke (ed.): *Bedeutende Frauen des 18. Jahrhunderts. 11 biographische Essays.* Würzburg: Königshausen & Neumann 2007, p. 162.

could become a true “field of experimentation”[16] for a social, political, cultural and literary situation in which the boundaries of class as well as national barriers, the boundaries between different religions and denominations as well as between the sexes became permeable. Forms of life and norms of life appeared as movable, as mobile.

In Rahel’s attic society, a Prussia emerged that could have been and yet was not allowed to be. A Prussia as a dream, a Prussia that never existed. It was a Prussia of potentialities, of polylogues across all borders, a Prussia for which the most brilliant minds gathered and wove threads of conversation that would reach far into the future. It was a country in which members of all classes, all origins and all religions could experience conviviality in difference, living together across all rifts and borders.

Certainly: this society lived a community as a dream for the future. The present soon became harsher. For the Napoleonic wars soon changed the political and social, but also the economic and cultural situation fundamentally and caused all hopes to fade of continuing such a conversation in its polyphony and radical openness and thus sketching out what could have been transferred from the imaginable to the thinkable and from the thinkable to the writeable, but also from the writeable to the liveable and lived. Rahel Levin’s salon, with all its figures of politicians and artists, writers and philosophers, men and women, formed a mobile in constant motion, in which relationships were forged between all its figures that could be regarded as configurations of a prospective Prussia: the Berlin salon as a mobile of an emerging, more open society.

The beating heart of this salon was one Rahel Levin, who was by no means only the master of the spoken word, of conversation and of stimulating and thoughtful talk. For she had begun very early on to build up a network of written communication alongside the all-important oral level, which she was to expand further in the years that followed. From her first pen pals, such as the Jewish medical student David Veit, who considerably broadened her educational horizons, she developed a complex network of relations, so that by the year of her death in 1833 an intensive correspondence had developed that probably comprised well over ten thousand letters.[17] It was a flow of information that did not cease even in the crisis year of 1808.

The thesis that Rachel was a ‘writer without an actual work’[18] has long been untenable. She had only written letters—so the argument went—and neglected the creation of an actual literary work. But such a view was based on a normative understanding of literature—the term only emerged historically in the last third of the eighteenth century in a form close to today’s understanding—which excluded or relegated to the margins precisely the so-called small forms such as letters or aphorisms.

---

[16] Ibid.

[17] Cf. Danzer, Gerhard: *Die Dachstubenwahrheiten der Rahel Varnhagen*, p. 59.

[18] Cf. Fuchs, Renata: ‘Soll ein Weib wohl Bücher schreiben; Oder soll sie’s lassen bleiben?’: The Immediate Reception of Rahel Levin Varnhagen as a Public Figure. In: *Neophilologus* (Budapest) XCVIII, 2 (April 2014), p. 314.

Rahel Levin, however, was a master of these small forms. For a long time Rahel Levin Varnhagen has been famous as an author in the true sense of the word, and the fact has been acknowledged that her correspondence is an essential part of a body of work that is also literary in the narrower sense.[19] Certainly she made use of more marginal[20] literary genres, and occupied herself chiefly with letters, diary entries, or aphorisms; but she did so brilliantly that here a true literary pioneer distinguished herself. For a long time the epistolary works of Rahel Levin Varnhagen have become canonical: The most diverse editions alternated and continue to alternate; her correspondence is more than just a testimony to her time and her being human. The history of its reception and impact will be discussed later; but it is a piece of high literature in the best sense of the word.

Rahel's pen pals are no less wide-ranging and varied than the visitors to her Berlin salon. On the one hand, there is the correspondence with her friends, in particular with Karl Gustav von Brinckmann, with Alexander von der Marwitz or with Pauline Wiesel,[21] with whom she remained in contact from her difficult years until her death.

But Rahel's correspondence extends far beyond exchanges with her intimate friends or lovers such as Count Karl von Finckenstein,[22] Wilhelm Bokelmann, Raphael d'Urquijo or Karl August Varnhagen von Ense. We see a letter writer at work who goes beyond the anecdotal or documentary, and in her correspondence with Wilhelm or Alexander von Humboldt, with Jean Paul,[23] Karoline Gräfin von Schlabrendorff or Heinrich Heine, not only openly advocates the emancipation of Jews in Prussia, but also the emancipation of women with verve.[24] The breadth of her themes, but at the same time their intellectual depth, is astonishing. This is true both in speech and in writing. Thus, for example, a Wilhelm von Humboldt notes retrospectively about the months they spent together in the French capital:

---

[19] Cf. Krauss, Andrea: Major Minor Form: Marginalia on Rahel Levin Varnhagen's 'Rahel. A Book of Remembrance for Her Friends'. In: *The German Quarterly* (Cherry Hill) 88 (Spring 2015), pp. 149–171; and generally Whittle, Ruth: *Gender, Canon and Literary History. The Changing Place of Nineteenth-Century German Women Writers.* Berlin/Boston: de Gruyter 2013.

[20] Cf. Krauss, Andrea: Major Minor Form: Marginalia on Rahel Levin Varnhagen's 'Rahel. Ein Buch des Andenkens für ihre Freunde', pp. 149–171.

[21] Cf. Schmid, Pia: Die Freundinnen Rahel Levin Varnhagen und Pauline Wiesel. Zum Freundschaftskult im deutschen Bildungsbürgertum. In: Labouvie, Eva (ed.): *Schwestern und Freundinnen. Zur Kulturgeschichte weiblicher Kommunikation.* Cologne/Weimar/Vienna: Böhlau 2009, pp. 101–119.

[22] Cf. Stern, Carola: *Der Text meines Herzens. Das Leben der Rahel Varnhagen.* Reinbek bei Hamburg: Rowohlt 1994, pp. 63–78.

[23] Cf. Albrecht, Andrea: Bildung und Ehe 'genialer Weiber'. Jean Paul's 'Diesjährige Nachlesung an die Dichtinnen' as a Response to Esther Gad and Rahel Levin Varnhagen. In: *Deutsche Vierteljahresschrift für Literaturwissenschaft und Geistesgeschichte* (Stuttgart) 80 (2006), S. 378–407.

[24] Cf. Albrecht, Andrea: *Bildung und Ehe 'genialer Weiber'. Jean Paul's ‚Diesjährige Nachlesung an die Dichtinnen'als Erwiderung auf Esther Gad und Rahel Levin Varnhagen*, pp. 378–407.

> When I was in Paris with my family, she [i.e. Rahel] was there for several months, and there was not a day that went by that we did not see each other. One was glad to visit her, not only because she was really of a very amiable character, but because one could almost be sure never to leave her without having heard something from her and having taken something away with one that gave one food for further serious, often deep thought or stimulated one's feelings.[25]

Just as Rahel Levin Varnhagen was a master of conversation and educated conversation, so she was a master of epistolary communication: leaping swiftly from one subject to another, moving back and forth in the minds of her contemporaries, imaginative, and always writing with that intensity which came so naturally to her in oral conversation as well as in written correspondence.

From her time in Paris comes a letter dated April 2, 1801 to David Veit, in which she expresses her assessment of the French capital and also her disappointment at a written entry that she forwards to her correspondent. The liveliness of her conversation is as present in these lines as the confrontation with an Other that had long since ceased to be an Other and thus a challenge for her:

> I wrote to myself the other day in a little booklet: 'Good things exist a long time before they have their reputation, and their reputation exists a long time when they are no more'. That is all I want to say to you about Paris. Long, methinks, is it, and can it no longer be Paris; after all Germany has become Paris for centuries. For Paris seems to me like a Germany huddled together, and little different. I could spin that out: some other time! do it yourself, in the meantime.[26]

In her letters, Rahel always aims to draw her correspondents into her game, to let them 'spin out' what she has found and invented in her dialogue with herself. Lively interaction is at the heart of her speech to her correspondents, without Rahel allowing it to fix her movement. The surprising turn is her basic trait, both orally and in writing.

In this respect, the moment of surprise is characteristic of Rahel, her attempt to let the other think with her, to write with her, and to draw her into her own game in a somewhat impish way. If the whole of Germany has become Paris, what else can be expected of a French capital that has long since embodied itself in its counterpart?

Rahel's insatiable curiosity and her continuing interest in all forms of expression of female existence are evident throughout, but especially in her correspondence with all those female correspondents who—such as the famous actress Auguste Henriette Brede[27]—sought to demonstrate to her the harsh professional conditions of a professional life as an independent woman. In essence, Rahel's correspondence

---

[25] Humboldt, Wilhelm von: Letter to Charlotte Diede (1834). In: Varnhagen, Rahel: *Gesammelte Werke*, vol. 10, S. 383 f.

[26] Varnhagen, Rahel: *Briefe und Aufzeichnungen.* Ed. by Dieter Bähtz. Frankfurt a. M.: Insel 1986, S. 66.

[27] See Deiulio, Laura: Performing German Women's Professional Identity: The Correspondence of Rahel Levin Varnhagen and Auguste Brede. In: *German Studies Review* XXXVIII, 3 (October 2015), S. 509–529.

is not a continuation of the salon in a different medium, does not spin on the oral salon conversation with other, written textual means. Rather, the correspondence opens up *another* world, a spatially and temporally extended world that goes beyond the here and now of immediate, direct interaction.

And it is the correspondence that becomes so important and weighty, especially in those years when her salon comes to a standstill and finally all her friends stay away from the attic room. One could say that Rahel Levin worked with a double net, as it were, and had already spun her second, written network long before her oral network was so abruptly severed by political events.

The mobile of the figures changes its configurations; yet it does not disappear, but continues to exist in altered written form before it is given a second chance in orality, experiences a second life, in a political environment that has again been fundamentally transformed.

Transitions

If the year 1808 also brought the final realization of the failure of all her plans to gather around her a small but select society that could be something like the microcosm of a future Berlin and Prussia, this year also heralded a development at the end of which there would be a new draft, a mobile of a Prussian society as it should exist, at least ideally. For it was in this year that Rahel Levin got to know Karl August Varnhagen better.

Rahel had already met the medical student, who was fourteen years younger, in 1803 and 1807,[28] but it was not until 1808 that the future couple became closer. From Varnhagen's side, there was very soon a passion for Rahel, but she had become more cautious in love matters and initially reacted wait-and-see. On May 12, 1808, their correspondence began, which already attempted to illuminate the entire scope of their relationship until their deaths. At a time when Rahel was separating from her mother and from the famous apartment in the Jägerstraße, and thus from her attic, a bond was formed between the two of them that was not to be severed, despite some turmoil and discord.

Certainly much has been written about this relationship since Hannah Arendt; and as a rule it was the younger Varnhagen who always came off very badly in all comparisons. As an example of this, let us just briefly mention the view of an editor of Rahel Levin Varnhagen's letters, who described the couple with the following words:

> Still, there was no mistaking it: the gifts were all too unequally distributed; all the glamour went out of her, he was little more than an admiring mirror.[29]

[28] Cf. Danzer, Gerhard: *Rahel Varnhagen—eine Frau entdeckt sich über die Kultur*, p. 161.

[29] Kemp, Friedhelm: Nachwort. In: Varnhagen, Rahel: *Briefwechsel.* Vol. II. Munich: Winkler 1979, p. 401.

When Friedhelm Kemp rightly notes that Hannah Arendt "discusses Rahel's Jewishness as a central theme of her life"[30] and aptly states that the volume is "in a very idiosyncratic way, and for that very reason, the most significant, independent, and forceful thing that has been written about Rahel to date,"[31] the figure of Varnhagen von Ense is also literally wiped away in Arendt's footsteps, indeed appearing only as a mere reflector of a light that was not his. But may we not assume that Hannah Arendt wanted to push this Varnhagen von Ense into the background because he could not be incorporated into the Jewish story of Rahel?

Be that as it may: Karl August Varnhagen von Ense stood by Rahel Levin—even and especially in her most difficult months and years. Last but not least, he also became the first editor of her letters and thus created the basis that still serves us today to recognize the whole greatness and significance of Rahel Levin Varnhagen beyond her activity as a salonière. And this man is supposed to have understood nothing, absolutely nothing of her? Nothing but a garrulous companion is he supposed to have been, or even a mute, if admiring mirror?

When Varnhagen von Ense joined the Austrian army in June 1809, immediately after the victory of the Austrians under Archduke Karl over Napoleon in May, and hoped, with good reasons and ultimately successfully, that he would be able to lay a better foundation for his life and advance in diplomacy, this soon marked the beginning of an unsteady period of travel and change of location, which, however, was not able to shake his relationship with Rahel in the long run. Thus the couple spent the summer of 1811 together in Teplitz and constantly expanded their circle of acquaintances and friends, so that a further network of relationships soon developed, which further nourished the network of correspondents.

The Austrian officer Varnhagen von Ense wisely ties up his connections and skillfully establishes links with Prussia that were to come in handy once more. But everything is still precarious, everything seems to hang by a thread: the future, politics, personal happiness, friends, indeed life and survival in wartime itself. Rahel writes from Dresden to Varnhagen von Ense on September 16, 1811:

> Hear for consolation that I grieve far more over being separated from you than I ever thought I would. I, too, feel quite anxious: I suddenly feel so torn away from protection, security, and love that I could walk all around me just to see, just to find: to whom do you belong? to what? Yesterday, towards evening, I took the most delightful walk with Marwitz and Lippe, probably a mile, up the Ostrawiese. You know whether and how I love Marwitz, they were two friends: we sometimes walked quietly, the wide space was great and divine, the splendid sun and evening glow, the serious and quite different trees from those in Bohemia, the endless avenues; only I had to think: you are alone and a stranger here, if these two do not want to go with you; alone and a stranger, even if they stay beside you; you are not their dearest, they do not relate everything to you. How surely I have lived till now! And I was not ungrateful, Varnhagen! do not take it so crudely as the word is here: it was not only

[30] Ibid., p. 403.

[31] Ibid.

> gratitude, it was loving longing; and my heart's longing answered yours, my heart kept time with yours. And so are most of my moments.[32]

In a time and context of uncertainty, in an era of great upheaval and unforeseeable change, Rahel notes the lack of protection, the lack of security, only to invoke the absence of the Varnhagen she lacks in the presence of new friends. Varnhagen von Ense, however, was always at pains to give Rahel the certainty and security she needed.

At the same time, he endeavours to publish Rahel's literary work, successfully asking Goethe in 1811 to publish excerpts from his correspondence with Rahel Levin: The correspondence appears in Cottas 'Morgenblatt' in 1812, to my knowledge the first publication of the young Varnhagen concerning Rahel's correspondence. At the same time, this not only gave expression to Rahel Levin's admiration for Goethe, but also laid the foundation for that exuberant Goethe cult that was to emanate from Rahel's second salon. But let us not anticipate.

During the turmoil of the Napoleonic Wars, the couple sees each other only rarely, but after the Battle of Leipzig and Napoleon's defeat, they spend the following summer of 1814 together again in Teplitz. For almost a decade, Rahel repeatedly follows her lover to various European cities and finally follows him to Vienna, where he participates in the Congress of Vienna in Hardenberg's service.

The time of travel, farewell and reunion is not yet drawing to a close, for the Napoleonic storm still reigns over Europe and has not yet abated. In private, things are turning around. In September 1814 Rahel Levin converts to Christianity and takes the baptismal name Antonie Friederike; her marriage to Karl August Varnhagen is soon to follow. Two weeks earlier, on September 13, 1814, she sends the following words to her Varnhagen:

> Golden August, how can I answer your affectionate, fine letter but with myself, as with the recognition of what you are […]. With me, my being right shows itself in your person: but that you are right is hard to see. This morning there was in my bed the Sopha cushion which you used in Teplitz. It stirred my whole heart with its tear, which it has from you! But now no more tender words! For—apart from time one can do nothing…[33]

Here again we see a Rachel as a letter-writer, tapping her pen pal, giving him an impulse, yet without saying and commenting on everything herself. The letter gives a good impression of the tone that prevails between the two, without revealing too much in each case—as if this letter was also conceived from the beginning in such a way that it could be published effortlessly.

Happiness arrives when Varnhagen is appointed Prussian chargé d'affaires at the Baden court in Karlsruhe in July 1816. For Rahel Varnhagen, after long years of travel, reunion and renewed separation from Varnhagen, these are, by her own

---

[32] Varnhagen, Rahel: Letter to Varnhagen von Ense, September 16, 1811, in (this): *Briefwechsel*, vol. 2, p. 166.

[33] Varnhagen, Rahel: Letter to Varnhagen of September 13, 1814. in (this): *Briefwechsel*, vol. 2, p. 295 f.

admission, the happiest years that she is allowed to spend in the Baden royal city. In October 1819, however, this time comes to an end. Varnhagen is recalled without giving any reasons: They move back to Berlin together. The time of transitions is over.

In the Salon of Rahel Varnhagen

Karl August Varnhagen, who soon had his title of noble von Ense acknowledged, developed after his recall as Prussian chargé d'affaires in Karlsruhe into the publicist, historian and collector as whom he is probably mostly known to us today. He later became famous for his correspondence with Alexander von Humboldt, which Rahel's niece Ludmilla Assing[34] published posthumously and, moreover, shortly after Alexander von Humboldt's death—not without thereby damaging Alexander's image, since one had not expected such openness and strong criticism of the Prussian royal court from him. But basically, the frankness with which Alexander von Humboldt allowed Karl August Varnhagen von Ense insight into his scientific and literary, but also his political convictions, was proof enough of the esteem in which Rahel's husband was held, at least in the liberal circles of Prussian society. The letters provide a good picture of both personalities and of their critical attitude towards those trends which Humboldt opposed for more than half his life. Varnhagen was an important interlocutor for him.

Soon the Varnhagen couple moved into a spacious apartment not far from Rahel's original sphere of activity. The written network of friends, confidants and companions still existed. Soon Antonie Friederike Varnhagen von Ense, who of course continued to be known to everyone as Rachel, was once again able to gather an illustrious circle around her, which is commonly referred to as her 'second salon'.

Fourteen years had passed since her attic room society had come to an abrupt end. In a changed political environment of the Restoration, this second circle also united members of different classes, scholars of the most diverse origins, Jewish writers and scholars in a circle that closed around the central figure of Rahel Varnhagen. For Rahel—and not her husband Varnhagen—formed the undisputed centre of their society, their 'salon': Her husband kept in the background, leaving the stage to his experienced wife, who knew how to attract nobles and commoners, Jews and Christians alike. Her community once again became a haven of free speech in a society that had become increasingly repressive after the Congress of Vienna and the Carlsbad resolutions. Other circles now set the tone, such as the Christlich-Deutsche Tischgesellschaft (Christian-German Society), co-founded by Clemens Brentano, which had fixed statutes and was reserved for men alone.[35]

---

[34] In this connection, cf. *letters from Alexander von Humboldt to Varnhagen von Ense from the years 1827 to 1858. Nebst Auszug aus Varnhagen's Tagebüchern, und Briefen von Varnhagen und anderen an Humboldt.* Ed. by Ludmilla Assing. Leipzig: F.A. Brockhaus 1860.

[35] Cf. Arendt, Hannah: *Rahel Varnhagen*, pp. 86 and 119 f.

In contrast to such circles of nationalistic romantics, many celebrities can also be cited with regard to Rahel's second circle, who populated her salon and lent it an impressive liveliness and openness. Among Rahel's guests were Prince Hermann von Pückler-Muskau or Alexander and Wilhelm von Humboldt, Leopold von Ranke or various members of the Mendelssohn family, Eduard Gans or Adelbert von Chamisso, Hegel, Fichte or Schleiermacher, the writers Heinrich Heine or Ludwig Börne, authors for whom Rahel was a strong advocate.

But the political environment had changed. Clemens Brentano's circle may be exemplary here, as he banished from his circle not only all Frenchmen, but also women, Philistines and above all Jews: his Christian-German Society propagated openly anti-Jewish and certainly also anti-Semitic tendencies, in which Brentano himself took an active part. The model for a future society here was openly restrictive and restorative; it stood in clear contrast to the conviviality and living together of the community that gathered around Rahel Varnhagen. This community, however, was now unmistakably minoritarian.

The changes between the first and second salons were nevertheless of a more gradual nature and can basically be described as follows:

> In her second salon, Rahel assumed not only the role of integration figure but also that of motherly educator and mentor. It is therefore no wonder that, in addition to acknowledged intellectual greats such as Hegel or Alexander von Humboldt, 'Young Germany' and the gifted offspring also regularly frequented her salon.[36]

In addition to artistic, literary, and philosophical subject areas, political topics now dominated in Rahel's second salon alongside the important cult of Goethe, with social utopian and especially Saint-Simonian ideas being of great importance. Some things seem like an escape into a future that had been postponed even further than in the first salon, which can be understood more as a model of society *en miniature* related to the present.

Perhaps this was a hint that a mobile of Prussia had been set in motion here, whose configurations would not take hold in an immediate present or near future, but rather attempted to anticipate and think ahead to something that might one day determine the constellations of a more distant future as a mobile model. But here, too, Rahel Varnhagen was able to translate the more political orientation of her salon into a mobile of possibilities that was to form both a model and a livable, indeed literally *lived*, reality in a small society. This prospective dimension of the second salon should not be overlooked.

In all her activities, Rahel Levin Varnhagen was now actively supported by her husband. Certainly he did not possess the literary and artistic radiance that was so characteristic of his wife; but he was always at pains to contribute positively as much to her network of oral dialogues and conversations as to promote her ever-expanding literary network. With not a few of the salon visitors he kept up an interesting correspondence on his part, and gained an overview of developments in

[36] Danzer, Gerhard: *Rahel Varnhagen—eine Frau entdeckt sich über die Kultur*, p. 164.

Prussia which continued to grow steadily in the quarter of a century after Rahel's death in 1833 until his death in 1858.

The deep grief over his wife's early passing is also evident in Varnhagen's correspondence, but very quickly gave way to his concrete efforts to honor Rahel's memory and to create and secure a place for her in the literary (and also political) world of Prussia. Only a few months after the death of Rahel Levin Varnhagen, he published a collection (still in agreement with her) whose importance simply cannot be overestimated: *Rahel. Ein Buch des Andenkens für ihre Freunde.*[37]

This collection, initially published in a single volume and subsequently in three volumes for the book trade, forms the origin, as it were, of all those editions and editorships that continue to accompany the various phases of Rahel Levin Varnhagen's reception to this day.[38] As already mentioned, it was only another of Varnhagen's editions of Rahel Levin Varnhagen's letters; but as a monographic work it was groundbreaking and forward-looking, since this collection was addressed to immediate friends, but consciously reached out far beyond the circle of friends to a general audience. It was the nucleus of that development which *post mortem* gave Rahel her status as a great German-language writer.

It is generally reported that at the beginning of the 1830s, both the salon guests and the pen pals noticed the slow waning of Rahel's strength. One suspected that she no longer had all her energy and intensity, that she was slightly ill and bedridden. Yet on 1 February 1833, barely more than a month before her death, Alexander von Humboldt wrote to her, curiously enough, quite concerned not for her but for Varnhagen von Ense. His choice of words in the letter to Rahel is remarkable: "Would you like to say a comforting word to me about the dear Varnhagen, the only brilliant pillar of our literature (in the nobler sense of the word) of our fatherland."[39] Such formulations express how much Humboldt appreciated Rahel's husband and how little he considered her own imminent death.

Five weeks later, however, Alexander von Humboldt learned of Rahel's death and addressed Varnhagen, who was grieving for her, in a letter dated 9 March of the same year:

> A spirit like yours, my noble friend, needs solitude and tranquility; you draw only from yourself. Think that I received the terrible news [of Rahel's death] only last night through

[37] Varnhagen von Ense, Karl August (ed.): *Rahel. Ein Buch des Andenkens für ihre Freunde.* 3 vols. Berlin 1834; I have the following edition: *Rahel. Ein Buch des Andenkens für ihre Freunde.* Berlin: Matthes & Seitz 2010.

[38] Cf. on the early reception history, among others, Volkening, Heide: Die Philologin, Rahel Varnhagen und das allgemein Menschliche. Varnhagen Studies in the First Third of the twentieth Century. In: Kauko, Miriam/Mieszkowski, Sylvia/Tischel, Alexandra (eds.): *Gendered Academia. Wissenschaft und Geschlechterdifferenz 1890–1945.* Göttingen: Wallstein 2005, pp. 237–253; and Whittle, Ruth: *Gender, Canon and Literary History. The Changing Place of Nineteenth-Century German Women Writers*, pp. 64–84.

[39] Humboldt, Alexander von: Letter to Rahel, February 1, 1833, in *Briefe von Alexander von Humboldt an Varnhagen von Ense*, p. 12.

> Prince Carolath. You know what a warm, long-suffering, thoughtful friend I am losing in her, the ornament of her family [...].[40]

Many of her friends reacted in the same way as Humboldt. The *book of remembrance* edited by Varnhagen *for her friends* was certainly intended primarily for the members of her circle, her salon. But it laid the foundation for her posthumous canonization and her rise, which continues to this day, to a central figure in the history of German literature and ideas.

With the selection of letters and diary entries made, it generated the literary work of Rahel Levin Varnhagen in the true sense of the word by bringing her writing within the reach of a broad public in the German-speaking world. Certainly, it took time before the real significance of Rahel Levin Varnhagen as a Jew and as a female writer, as a fighter for the emancipation of Jews as well as for the emancipation of women, could be understood. But a beginning had been made, the foundation laid.

Not only her oral, polylogical work, but also her written creations were now before everyone's eyes and testified to a writer who was well aware of her doubly marginalized role as a Jew and as a woman and who, with her conversational as well as her correspondence partners, openly advocated a double equality—that of Jews and that of women—in the Prussia of the time.

A few years later, Heinrich Heine, in his *Buch der Lieder*—an obvious title tribute to Rahel's memory—judged in the following way:

> It is as if Rahel knew what posthumous mission was in store for her. She believed, of course, that things would get better, and waited; but when there was no end to the waiting, she shook her head impatiently, looked at Varnhagen, and died quickly—in order to be able to rise again all the more quickly.[41]

Heinrich Heine not only recognized in all clarity that nothing was moving in Prussia at the time in the very sense that drove the small society in Rahel Levin Varnhagen's two salons and thus her Mobile Preußen. He also recognized very sharply that Rahel's literary work still had its future before it, that her 'mission' was by no means over. For Rahel Levin Varnhagen's impact unfolded far beyond her death—and also far beyond the decades of political stagnation in Prussia.

---

[40] Humboldt, Alexander von: Brief an Varnhagen von Ense vom 9. März 1833. In: *Briefe von Alexander von Humboldt an Varnhagen von Ense*, p. 16.

[41] Heine, Heinrich: Vorrede zum Buch der Lieder (1837). In: Heine, Heinrich: *Sämtliche Schriften*. Ed. by Klaus Briegeleb. Vol. 1. Munich: Carl Hanser 1968, p. 10. Cf. on this Danzer, Gerhard: *Rahel Varnhagen—eine Frau entdeckt sich über die Kultur*, p. 165.

Sociability, Communication and Conviviality

If one considers the life of Rahel Levin Varnhagen in its entirety, it is clearly marked by a sociability that was certainly passed on to her from her father.[42] Her entire life is inscribed in the development of new forms of this sociability,[43] of friendship and intimacy, as well as of a conviviality that emerged in new manners at the turn of the nineteenth century in Prussia.

For Jewish women and their emancipation in families where a strict patriarchy often prevailed, it was of the utmost importance to achieve a playing field in the founding of regular meetings and salons that made them widely visible in the society of the time and opened up new forms of living together, which until then had been characterized by the rigid barriers of an essentially outdated class order. The socio-cultural conditions for the young Rahel Levin were therefore favorable.

Even if the social framework conditions had changed during the period of transition between the two salons, the structures that had once been created, and above all their network of correspondents, proved to be so persistent that a certain continuity seemed to be guaranteed under changed political auspices. This may explain why it was possible for a second salon to form in the first place.

In this respect, the salon, the attic society, offered Rahel Levin, like Rahel Varnhagen, the invaluable opportunity to escape from traditional role expectations and to avoid being quickly married off by her brothers (like her sister Rose) after the death of their father. New forms of sociability, as demonstrated by the emerging salons of Jewish women, were the decisive means of disrupting not only outmoded social but also gender orders.

And new forms of intimacy, of friendship and love offered modes of self-realization that seemed highly attractive, especially for Jewish women, who were doubly marginalized in Prussian society. Rahel made lively use of them.

If Germaine de Staël[44] had the impression during her visit to the German salons that it was possible to communicate much more non-hierarchically here than in France between the classes and even to come into contact easily with the royal family, this observation could also be transferred to the norms such as the forms of gender relations and above all also the conceptions of love. For Rahel, this offered a wide field for experimentation, which she very consciously knew how to use for herself. For she liked to experiment.

When Hannah Arendt saw Rahel Levin Varnhagen's development from parvenue to pariah stretching out[45] she gave—not without autobiographical emphasis—great significance to the field of tension between the two poles, especially with regard to the situation of Jewish women in Berlin society. Perhaps the concepts of parvenue and pariah mark quite well the playing field that remained for marginalized Jewish

---

[42] Cf. Danzer, Gerhard: *Die Dachstubenwahrheiten der Rahel Varnhagen*, p. 59.

[43] Cf. Benhabib, Seyla: *The Pariah and Her Shadow: Hannah Arendt's Biography of Rahel Varnhagen*, p. 94.

[44] Cf. ibid., p. 97.

[45] Cf. on this Ette, Ottmar: *Die Aufgabe der Philologie*, p. 185.

women to increase their visibility and develop non-traditional ways of life. In any case, they were offered the opportunity to take their fate courageously into their own hands and shape their lives.

Like no other, however, Rahel Levin Varnhagen was able not only to explore this scope, but also to make use of it for herself and at the same time to gain a high national and international reputation. If, on the basis of the reception testimonies as preserved in the Varnhagen estate of the Jagiellonian Library in Cracow, Rahel could be described in a recent study as a "contemporary superstar",[46] this gives an insight into the internationality of the Berlin Jewess's reception and impact history and an impression of how she was already perceived in part by her contemporaries. Rahel Levin Varnhagen was by no means seen merely as a passing focal point of a salon she led: In France in particular, she was also credited with important functions and abilities in the history of ideas and philosophy.

If Rahel could consequently be perceived as a major figure within the German-language literary scene at the turn of the century and far beyond, this is inconceivable without her written testimonies, without her diary entries and aphorisms, but above all also her letters, which remain fascinating to this day. The large number of very different correspondents shows that, on the one hand, she understood the letters in a certain way as a kind of extension of her activity as a salonière and wanted to continue and prolong the oral conversation with a personality of her salon now in written form. At the same time, however, it also and above all shows that she cultivated a second circle of friends and acquaintances in written form, which allowed her to go far beyond the spatial and temporal limits of immediate face-to-face communication and thus exert an enormous influence on the literary-artistic scene of her time.

At the heart of Rahel Levin Varnhagen's life and life's work was simply communication in all its various forms. It was this that drove and inspired her life time and again: it gave it meaning. The most diverse forms of sociability that she cultivated in her two salons were linked to that lively, impulsive correspondence, rich in surprising turns of phrase, which can certainly give us a sense of the liveliness of her oral conversation both within the salon and outside it.

Conversation and correspondence formed the pillars of a communication that transcended the boundaries of class, profession, and religion and aimed at a community that possessed an unquestionable political claim in both salons, albeit with varying degrees of social realizability. This included body politics—such as the question of the fate of raped women, which was discussed at the time—as well as political questions of religious identity formation or access to education for broad sections of the population. In a broadly understood sense, politics was their domain, not only with regard to a saint-simonist world of ideas.

Thus we can understand her two salons, the Salon of Rahel Levin as well as the Salon of Rahel Varnhagen, not only as drafts of a new sociability, but as projections

[46] Cf. Fuchs, Renata: *‚Soll ein Weib wohl Bücher schreiben; Oder soll sie's lassen bleiben?': The Immediate Reception of Rahel Levin Varnhagen as a Public Figure*, p. 303. The phrase "contemporary superstar" was used in order to push Rahel's impact in foreign countries (p. 309).

and projects of a certainly small, select society that was designed for community. Their goal was a living together, a fundamental conviviality, which in its core was to lay the foundations of a new, a different society—a society in constant exchange, in constant interaction, a society as a community of those who are connected and think together.

In this way, and with the innovative means of her epoch, some of which shocked her contemporaries, Rahel Levin Varnhagen tested a mobile Prussia in which her various figures became components of those configurations that designed other forms and norms of conviviality, coexistence and participation for Prussia. It was a dreamed[47] Prussia, to be sure: the Prussia of an integration and conviviality in difference, transverse to the common images of identity assignments for individual social groups. Its model of society was relational and integrative: it did not rely on exclusion, but on inclusion of different social or religious groups. In this sense, it was clearly ahead of its time.

It may well be that her project failed, that her field of experimentation ultimately had to be abandoned in the turmoil of a society that set out on the path of nationalism, anti-Judaism and anti-Semitism, of class privileges and gender hierarchies. So it may be that Rahel Levin Varnhagen stands for a buried tradition of Prussian and German history that has still not lost its potential.

But her correspondence, above all, still conveys to us much of that conviviality that Rahel dreamed of between parvenue and pariah, a conviviality based on an all-encompassing communication in which all would be connected to all and enter into a fruitful polylogue with one another. The society she dreamed of may have been a distant dream, but her community was not.

[47] See Hahn, Barbara: *'Im Schlaf bin ich wacher.' Die Träume der Rahel Levin Varnhagen.*

Wilhelm and Alexander von Humboldt or Humboldtian Science

'Everything Is Interaction'

"*Everything is interaction*."[1] With these few words, which in the midst of a passage written in French detonate in German almost like a formula, Alexander von Humboldt in his *Amerikanischen Reisetagebüchern* summed up the basic axiom of the science named after him today as Humboldtian science.[2] It was no coincidence that it was Wilhelm von Humboldt who was probably the first to recognize his brother's fundamental model of thought, which the latter was to develop and outline a few years later on his great journey through the tropics of the American continent (1799–1804). For already in 1793, in a letter to Karl Gustav von Brinkmann, Wilhelm had attested to his younger brother a special gift, a skill in combinatorics, and described him as predestined to "connect ideas, to see chains of things that would have remained undiscovered for ages without him".[3]

With his reflections, Wilhelm advanced to become the first exegete, indeed the first epistemologist of that Humboldtian Science that was only beginning to emerge in 1793 and that certainly unfolds less in chains than in a net-like fabric. In the first volume of his *Cosmos*, dated Potsdam in November 1844, Alexander himself spoke of a "general concatenation, not in a simple linear direction, but in a web-like intertwined fabric".[4] We may add today that this model of thinking, science and writing is of a profoundly relational nature and does not align itself according to simple

[1] Humboldt, Alexander von: *Amerikanische Reisetagebücher*, vol. IX, pp. 27r, 2–27v.

[2] The term was introduced into research by Cannon, Susan Faye: *Science in Culture: The Early Victorian Period*. New York: Dawson and Science History Publications 1978, pp. 73–110, and further developed by Ette, Ottmar: *Alexander von Humboldt und die Globalisierung. Das Mobile des Wissens*. Frankfurt a. M./Leipzig: Insel 2009, pp. 16–22.

[3] Humboldt, Wilhelm von: *Briefe an Karl Gustav von Brinkmann*. Ed. by Albert Leitzmann. Leipzig 1939, p. 60. authoritative on Wilhelm von Humboldt still Borsche, Tilman: *Wilhelm von Humboldt*. Munich: Beck 1990.

[4] Humboldt, Alexander von: *Cosmos. Kosmos. Entwurf einer physischen Weltbeschreibung*. 5 vols. Stuttgart/Tübingen: Cotta 1845–1862, here vol. I, p. 33.

O. Ette, *Mobile Prussia*, https://doi.org/10.1007/978-3-476-05839-3_7

causal chains, but seeks to take into account a wide variety of factors. Rather, it is concerned with multi-logical fields of factors that interact with each other and correspond entirely to Humboldt's basic axiom, according to which everything is to be understood and examined as an interaction.

But how would it be if we understood Wilhelm von Humboldt not only as the first epistemologist of the Humboldtian Science that did not yet exist in 1793, but that could at best only be imagined in *nuce*? How, then, if Wilhelm were part of this science not only concocted by Alexander, but conceived and associated with his name? If, in other words, there were a science, a *Humboldtian* Science, that could be associated with the name of both brothers?

Panoramic studies are not generally suspected of necessarily expressing risky, perhaps even radical, or root-and-branch thinking, or of revolutionizing previous classifications. The current context, however, could offer a favorable opportunity for this. And opportunity should also and especially in science inspire love to go beyond existing demarcations.

Ever since I first became involved as a young man with the work of the great Cuban poet, essayist and revolutionary José Martí, I have been fascinated by a little confusion in the thinking of this late nineteenth-century thinker and globalization theorist, who resided for many years in Manhattan and was internationally well-informed. For him, as for most Latin Americans, the name Humboldt was 'naturally' associated with that of the scholar who travelled widely throughout the Spanish colonial empire in the Americas, and whose writings revealed something like the birth certificate of an America independent of Iberian colonialism. José Martí, however, thought, at least for a time in his exile in Venezuela, the homeland of Simón Bolívar, that this man who had done so much for the 'New World' was none other than the scholar and explorer of the Basque language (which the young Martí somewhat unhappily referred to as 'dialecto') of whom he had also heard and read.[5]

It has always seemed to me a worthwhile enterprise to understand this little confusion as a productive *misreading* and to use it for lines of thought that might open onto something unheard of, or at least hitherto unheard of. I do not mean that Wilhelm and Alexander were one and the same person who had used two different *noms de plume* in order to be able to work on different fields of knowledge in this way—an idea that would nevertheless possess much that is tempting if a gifted author were to play it out in literary form. What is meant, rather, is that from this point Alexander and Wilhelm could be thought together in such a way that this would not result in a simple homogeneous unity, but rather in the unity of the multi-logical and, at the same time, in a comprehensive sense—and please forgive the

[5] See, among others, the reflections in Martí's 'Sección Constante', which appeared in *La Opinión Nacional* on 14 January 1882, in Martí, José: *Obras Completas.* Vol. 23: *Periodismo diverso.* La Habana: Editorial de Ciencias Sociales 1975, p. 152.

sound play—a multi-logical philology[6] that would be able to pursue different logics simultaneously.

This has so far been countered by the fact that for long periods of time research on Wilhelm and that on Alexander von Humboldt were, and to some extent probably still are, separated by clear, sometimes even hostile demarcations. When I gave a lecture on Alexander von Humboldt at the Jagiellonian University in Krakow in early June 2017 as part of a very successful Polish-German collaboration, I was introduced right at the beginning by a specialist on Wilhelm von Humboldt based in Germany—I hope this was done in jest—with the words that I was "from the opposite party". Scholars who work on Wilhelm or Alexander von Humboldt—let us jokingly call them 'Wilhelmists' and 'Alexandrians'—have mostly experienced a rather different academic socialization, come from different and not infrequently divorced disciplines and usually specialize in one *or* the other of the two Humboldt brothers. Researchers who work in a balanced, even distribution across Alexander *and* Wilhelm, on the other hand, are rare (to put it mildly). One devotes oneself—to call up the most common clichés—either to the Prussian patriot or the cosmopolite, either to the linguist or the naturalist.

Within the long history of Humboldt research—and by this I mean both the *Humboldt Industry* devoted to Wilhelm and the *Humboldt Industry devoted* to Alexander—it is certainly possible to distinguish, somewhat simplistically, between three phases. In the first phase, the *spiritus rector* of the Friedrich Wilhelm University, which was given the name of both brothers after its reopening in 1949, and the world traveller had little in common. This also applies to the iconography of both brothers, which follows very different lines of tradition. The depiction of both brothers in a single painting, in a single picture, is therefore very rare and, as far as I know, only occurs when—as in the famous contemporary engraving by Friedrich Schiller, the mutually connected pair of brothers Humboldt and Johann Wolfgang Goethe in Jena—transcending connections open up.

Even the two statues of the brothers have a very different history of origin and foundation and follow different traditions of representation. It is as if Wilhelm, born in Potsdam in 1767, and Alexander, born in Berlin in 1769, whose two hundred and fiftieth birthday we are celebrating this year, had belonged to different worlds for a long time; one an inhabitant of the 'Old', the other of the 'New World'. But is there no relationship between the two worlds?

In a second phase of Humboldt research and Humboldt reception, the brothers were even stylized as antipodes, insofar as one became the epitome of the humanities scholar and the other the icon of the natural sciences. In this way, the notorious invention of Charles Percy Snow's thesis of the *Two Cultures* (which is undoubtedly based on a significant prehistory), which had long since become fragile but still

[6] On the outline of aiellogical philology, see Ette, Ottmar: *Viellogische Philologie. Die Literaturen der Welt und das Beispiel einer transarealen peruanischen Literatur.* Berlin: Walter Frey—edition tranvía 2013.

continues to have an effect, was[7] anchored, as it were, within the family and thus served an image of a division between 'the humanities scholar' Wilhelm and 'the natural scientist' Alexander that is obviously still omnipresent in a broader public, but not infrequently even in the academic world. However, in the light of Snow's two-cultures thesis, this divorce, which for a long time was a cheap one, is just as much a thing of the past as the no less false, but in Germany apparently ineradicable thesis that Alexander von Humboldt was 'one' or even 'the last universal scholar'.

It seems to me that today more than ever it is necessary to replace such outdated and backward-looking commonplaces with a prospective understanding of science that understands Humboldtian science, first developed on the basis of Alexander's example, as a conception of science oriented towards the future, whose present and future potential is far from exhausted.

For some years now, there have been increasing indications that a third phase in Humboldt research devoted to the two brothers seems to be opening up, in which the idea of separate worlds and the antipodal, ultimately bipolar structure of the pair of brothers and of the sciences in general are being overcome. In the year of Wilhelm von Humboldt's birth, we are offered a perhaps historical opportunity not only to *deal with* the two brothers separately, but to *think about* them *together* out of their clear but fruitfully bridgeable difference. For many of the developments in Humboldt's science have, to use a famous formula of Snow's, "the future in their bones".[8]

The Languages of Science

When the question of a Humboldtian science is posed in the following, which refers to both brothers and whose epistemology was recognized early on by Wilhelm in the case of Alexander, then this is by no means intended to conceal the differences, indeed the sometimes profound differences between the two brothers. In both, however, there is a common basis, which is based on the conviction that science has a public responsibility and aims at a democratization of this public sphere that must be continued unceasingly. Science means responsibility, today more than ever. For a science that does not transfer its knowledge to society fails to recognize its social responsibility and in the end is itself to blame if society deprives it of its resources.

Both Wilhelm's and Alexander's work aimed to have a social impact, indeed at times a socially explosive impact, since they sought to make their ideas and conceptions socially acceptable not only at the Prussian or French, at the Spanish or Habsburg court, not only in Parisian or Berlin salons, but also among a broad public in Europe as well as in America. Their model of knowledge and science, education and training is ethically founded and aims at a change not only of the contents of knowledge or the objects of research, not only of the structures of educational as

[7] Snow, C.P.: *The Two Cultures.* With Introduction by Stefan Collini. Cambridge: Cambridge University Press 1993.

[8] Snow, C.P.: *The Two Cultures*, p. 10.

well as scientific institutions, but at a social change in the sense of a democratization of knowledge, as Wilhelm's drafts for the re-foundation of a university or the reformation of our academy as well as the sixty-one *Cosmos Lectures* and even more the sixteen *Cosmos Lectures* of Alexander in the Singakademie, which were also accessible to women, may show. For this it was necessary to find and invent a language and, better still, suitable languages in memoranda and lectures, letters and books. Science is also translation.

Education and training, knowledge and science-related socialization of the brother and sister show weighty parallels, of which only a few are mentioned here. They learned under and with the best tutors Berlin had to offer, so that Schloss Tegel became their first and much pre-programmed place of education. I expressly do not endorse the widespread devaluation, indeed a veritable mobbing of their mother, which runs through almost all recent publications on the Humboldt brothers. Marie Elisabeth von Humboldt, who came from a Huguenot family and was born in Colomb, was the one who pulled all the strings after the early death of their father. She very consciously and energetically created the opportunities that the two brothers then used in their own highly successful way. It was not without reason that Wilhelm remarked in a letter how much the death of his mother had affected him, as if his own happiness had been carried to the grave with her death.[9] Both studied at the Viadrina on the "frosty shores of the Oder",[10] both quickly moved on to the then excellent University of Göttingen, both were impressed and influenced[11]—albeit in different ways—by Georg Forster, who had sailed around the world with James Cook, and yet went beyond him; indeed, both favored the same Bushman in the last phase of their lives, who was of such importance in the composition of their writings. Shared biographies make a double mark.

In the lives of both brothers, languages played an overriding role from the very beginning. If we look at the 2017/18 annual theme of the Berlin-Brandenburgischen Akademie der Wissenschaften, 'Language', and the "diversity of world references" and the "linguistic constitution of our living world"[12] it quickly becomes apparent in what a comprehensive sense the lives, works and activities of the Humboldt brothers are marked by a diversity of languages.

Both brothers spoke or communicated, wrote or published beyond their German and French language competence not only in a multitude of European languages

[9] Cf. Humboldt, Wilhelm von: Materialien. First volume: 1797–1798. Sunday 31st December (11th Nivôse n.st.). In: *Wilhelm von Humboldts Gesammelte Schriften. Tagebücher.* Edited by Albert Leitzmann. First volume. Berlin: de Gruyter 1968, p. 390.

[10] Humboldt, Alexander von: *Die Jugendbriefe 1787–1799.* ed. and explained by Ilse Jahn and Fritz G. Lange. Berlin: Akademie-Verlag 1973, p. 4.

[11] Cf. inter alia Schmitter, Peter: Zur Wissenschaftskonzeption Georg Forsters und dessen biographischen Bezuegen zu den Bruedern Humboldt. A preliminary study on the relationship between 'general natural history', 'physical description of the world' and 'general linguistics'. In: Naumann, Bernd/Plank, Frans/Hofbauer, Gottfried/Hoojkaas, Reijer (eds.): *Language and Earth: Elective Affinities between the Emerging Sciences of Linguistics and of Geology.* Amsterdam: Benjamins 1992, pp. 91–124, esp. 92.

[12] Annual theme 2017/2018 'Language': jahresthema.bbaw.de

and had profound skills, including their own translations, in the languages of occidental antiquity, but also studied and used non-European languages of antiquity and the present. In doing so, they were highly aware of the linguistic constitution of all knowledge and thus also of all sciences.

In addition, they used—apart from their distinctive diplomatic and political linguistic skills—a variety of scientific languages, as they had developed after the end of Latin as the dominant scholarly language in the most diverse vernacular languages, and worked themselves in disciplinary highly specialized languages, between which—as their writings show—they were able to switch back and forth and to translate. In doing so, both not only—as in the case of navigation, for example—sketched its history and stories, but also exerted a formative influence on the linguistic development and shaping of the most diverse disciplines and scientific languages. In an almost ideal-typical way, they embody the idea of multilingualism in the sciences.

If today, in the shadow of a still growing dominance of English, we are moving towards a dangerous monolingualism, in which basically entire libraries and archives, entire bodies of knowledge and forms of thought of the non-English realm are threatening to disappear, it could be[13] stressed, quite in the spirit of the recent plea by Jürgen Mittelstraß, Jürgen Trabant and Peter Fröhlicher in favour of a multilingualism in the sciences, that the linguistic diversity of the Humboldt brothers could not stand for an expiring model of science, but rather for a development of science for which it is worth fighting today more than ever, at least in the Humanities and Cultural Studies (and this translation alone shows how different the respective accentuations can be). This multilingualism of the two Humboldts, only briefly sketched here, is a prerequisite for their wealth of thought and knowledge. They never—not even in the political sphere—fell for a single language.

Whatever the future of the language(s) of science on our planet may look like: We will only be able to escape a significant reduction of—to quote the annual theme of the Berlin-Brandenburg Academy of Sciences and Humanities once again—the "diversity of world references" as well as the "linguistic constitution of our living world" if we do not release the ideal of scientific multilingualism for occasional disposal. The predominance of English does not mean the end, but it does mean an irretrievable loss and a substantial reduction, especially in the humanities and cultural studies. The actual model for the future—at least for these sciences—is rather the multilingualism of Wilhelm and Alexander von Humboldt, which is both principled and structural.

Consequently, one encounters languages about languages when one engages with the Humboldt brothers and the practice of Humboldtian scholarship that transcends both languages and disciplines. It is precisely the forms and norms of translation as well as the concrete translational practices that play a decisive epistemological role in Alexander's *American Travel Diaries* as well as in the manuscripts of Wilhelm's *Kawi work*. Again and again, the serious play as well as the

[13] Mittelstraß, Jürgen/Trabant, Jürgen/Fröhlicher, Peter (eds.): *Wissenschaftssprache. Ein Plädoyer für Mehrsprachigkeit in der Wissenschaft.* Stuttgart: J. B. Metzler 2016.

playful earnestness of the two brothers can be discerned in their dealings with languages as well as with their fields of activity. On the linguistic as well as on the meta-linguistic level, on the level of language use as well as on the level of language description, for example, there is less a juxtaposition than a coexistence and above all a confusion of the most diverse languages.

This multilingual juxtaposition, interlingual intermingling, and translingual confusion, in the incessant linguistic penetration of the counter-state of language, point to the fact that a living coexistence, a dynamic conviviality of languages, is brought before our eyes here. With regard to this conviviality, we could speak of an ecology of the polylogical, since what is at stake here is the interaction between different kinds of logics for which the languages stand. But such a conviviality of languages, on the basis of incessant translation activity, is the necessary condition for a conviviality of human beings, and perhaps even more for a coexistence of humanity.

The design of this future model of a Humboldtian Science, which had been emerging since the 1790s, was based on the linguistic constitution of all knowledge and was not limited to the realm of science, either on the level of the philosophy of language or on the epistemological or poetological level. For Wilhelm, like Alexander von Humboldt, was convinced that the world—this is also shown by Wilhelm's concept of the 'world view'—cannot be adequately grasped and understood in its complexity from the standpoint of a single language. Without a multilingual approach to the most diverse objects, a multi-logical understanding of the world in its diversity could not and cannot be achieved. The linguistic constitution of Humboldtian science thus represents a fundamental challenge not only for the past future of the nineteenth century, but above all for our future.

Commonality of Concepts, Commonality of Understanding

The dissimilarity of the two brothers in their character, but also in their lifestyle—differences that can be clearly seen in their iconography—has repeatedly led to misunderstandings and disagreements between them. Thus Wilhelm, then Prussian envoy in London and 'deported' abroad by his opponents in the Prussia of the post-Napoleonic reaction, reported on December 3, 1817 to his wife Caroline on the occasion of his brother's visit to the English capital that Alexander, in addition to "personal affection", always brought "movement into life".[14] "But true it always remains that one is sorry how he has ceased to be German and has become Parisian in every little thing."[15] He added, not without admiration, that Alexander was certainly "one of the strangest natures that ever existed";[16] but he, who, like his wife Li, felt himself the representative of a liberal patriotic Prussianism, and who had

[14] *Gespräche Alexander von Humboldts.* Ed. by Hanno Beck. Berlin: Akademie-Verlag 1959, p. 51.

[15] Ibid.

[16] Ibid., p. 52.

recently been called an "unconventional liberal,"[17] was as much disturbed by the worldly manners of his brother, long since transformed into an international celebrity, as he had already been offended by the latter's conduct during the Napoleonic occupation of Prussia, when Alexander, faithful to his science committed to humanity, had remained in Paris. For any nationalistic discourse, Alexander was unattainable throughout his life, just as after his death, even under the National Socialists, any attempt to instrumentalize him for nationalistic and even worse purposes had to fail.

Contrary to all the legends to the contrary, however, this did not change the brothers' deep mutual affection. No, neither brother fed the other rat poison, as Daniel Kehlmann wittily invented in an almost forgotten bestseller, just as Alexander—a fairy tale that began to spread at times—in no way paid homage to pedophile practices.[18] However much they differed in habitus, language, and way of life, as well as in some political assessments: Throughout their lives there is ample evidence of a great familiarity, affection and tenderness, as well as mutual admiration for each other's activities and personalities.[19]

When Alexander set out on a still uncertain journey around the world and said goodbye to Paris, his last glances were at his beloved brother: "I looked at Bonpland, with whom I was to undertake such a long journey. What a marriage! The diligence continued. My eyes watched William the longest. He looked very cheerful, and that did me infinite good." With him, among his measuring instruments, his faithful traveling companions, he carried a "Renard's little thermometer, which Wilhelm had had so long (it is therefore so dear to me)."[20] And when he finally arrived in his longed-for tropics, he immediately wrote an enthusiastic letter to his brother Wilhelm from present-day Venezuela on July 16, 1799,[21] which can serve, as it were, as the key to the entire American voyage. Later, too, he sent his brother a large number of expressive letters, in which, incidentally, he liked to emphasize the dangerous nature of his 'adventures' in the tropics.[22]

Yet the two brothers were close not only in their feelings but also in their writings. After Wilhelm's death, Alexander took careful care to edit his brother's writings; and conversely, long years before, he had appointed Wilhelm to edit a central part of his own records from America, adding confidently that Wilhelm would certainly correct the style of his works.[23] Each knew the other's work, each appreciated its value: Throughout their intellectual biography, both cultivated mutual intellectual exchange.

---

[17] Maurer, Michael: The Unconventional Wilhelm von Humboldt. A commemoration of his 250th birthday. In: *Forschung & Lehre* (Bonn) XXIV, 6 (2017), pp. 490–493, here p. 493.

[18] Kehlmann, Daniel: *Die Vermessung der Welt. Roman*. Reinbek bei Hamburg: Rowohlt 2005.

[19] Humboldt, Alexander von: *Amerikanische Reisetagebücher*, Bd. II und VI, 52v 2.

[20] Ebd., Bd. I, 10r–10v 3.

[21] Alexander's letter to Wilhelm from Cumaná on 16.07.1799 in Humboldt, Alexander von: *Briefe aus Amerika 1799–1804*. Edited by Ulrike Moheit. Berlin: Akademie-Verlag 1993, pp. 41–43.

[22] Cf. ibid. especially letters Nos. 7, 11, 18, 35, 43, 51, 55, 62, 71, 79, 89, 111, 115, 134 and 142.

[23] Humboldt, Alexander von: *Amerikanische Reisetagebücher*, vol. VIII, 167v.

Thus the Humboldt brothers—and this is often lost in the specialized individual accounts—worked together intensively in a wide variety of fields even during both of their lifetimes. But did they also have a common project, a research project shared by both of them?

Such an idea emerges from a renewed look back to the year 1793, in which Wilhelm not only advanced to become Alexander's epistemologist, but also formulated, with explicit reference to his brother, what may very well be understood as the common research program of both brothers.[24] In his letter of March 18, 1793 to Karl Gustav von Brinkmann, Wilhelm expressed it, not without pathos, as follows:

> A complete restoration of the sciences and, more than this, of all human endeavor has been necessary for centuries, and the necessity grows with every passing year [...]. The most important step towards this restoration is to bring unity into all human striving, to show that this unity is the human being, namely the inner human being, and to portray the human being as he acts on everything outside of him and as everything outside of him acts on him, to draw from this the state of the human race, to sketch its possible revolutions and to explain the real ones as much as possible.[25]

This certainly anything but modest program for the "Restoration of the Sciences" aims in anthropological orientation at the unity of man and perhaps even more of the human race on the basis not of a homogeneous and static, but of a dynamic understanding of what is to be at the center of these Natural *and* Cultural Sciences, characterized by constant changes and revolutions. Differences of orientation, but not a divorce into those *two cultures,* which were later to be so persistently and obsessively projected onto the science of the Humboldt brothers, are indeed taken into account here.

The formula "how he affects everything except him and how everything except him affects him", which from today's perspective has an unmistakably ecological effect and is based on interrelationships, cannot be split into a 'humanities' and a 'natural sciences' part, but rather opens up to that unifying and binding formula that Alexander found in his *American Travel Diaries*: *Everything is Interaction.* The Humboldts are concerned with these interactions between man and his living world, between man and his environment, which has long since been changed by him, in a reciprocal network of relationships that is to be explored scientifically by both. Alexander's *avant la lettre* geoecological schemes of thought are in this early tradition.

The interactions between the brothers could, there can be no doubt, fuel an entire academy project and more. Only a few aspects can be mentioned here beyond what has already been mentioned. Perhaps foremost among these is the view, shared by the two brothers, of science as an open-ended process that can never be completed.

---

[24] Cf. on this Schmitter, Peter: *Zur Wissenschaftskonzeption Georg Forsters und dessen biographischen Bezuegen zu den Bruedern Humboldt. Eine Vorstudie zum Verhaeltnis von 'allgemeiner Naturgeschichte', 'physischer Weltbeschreibung' und 'allgemeiner Sprachkunde'*, p. 117 f.

[25] Humboldt, Wilhelm von: Brief an Karl Gustav von Brinkmann vom 18.03.1797. In Wilhelm von Humboldt: *Briefe.* Vol. 2: Bd. 2: Juli 1791–Juni 1798. Edited and commented by Phillip Mattson. Berlin: de Gruyter 2015, p. 141.

What is true for both writings is that they are characterized by a fundamental, at times almost radical incompleteness, which always distinguishes them as a growing *work in progress.* Even if Alexander had almost a quarter of a century more time at his disposal for research, thinking and writing, this basic trait of incompleteness by no means applies only to Wilhelm, but also to his younger brother: everything is in motion in the Humboldts' work, opening up to a future for which they seek to prepare the way. The fundamental incompleteness of their works is not epidemic, but epistemic.

In a series of lectures and essays, Jürgen Trabant has drawn attention to the Humboldts' close collaboration in the field of linguistics and, in a recent publication, marked August 3, 1804, i.e. the day Aimé Bonpland and Alexander von Humboldt returned to Europe, as a particularly significant day in the history of science: For this had been "after Leibniz's birth certificate of 1765, the second hour of the birth of European linguistics".[26] Without being able to go into detail about the results of this study at this point, it should be noted that cooperation in the field of linguistics was by no means limited to Alexander's collecting and obtaining grammars of indigenous languages, but was of a substantial nature.

The materials selected and brought by Alexander would have triggered "a complete reorientation of *historical-comparative* linguistics" as well as a "post-Leibnizian re-foundation of *anthropological-comparative* linguistics"[27] through the cooperation that developed from it, and would have induced Wilhelm to write his book on the American languages, which was admittedly never completed, for Alexander's travel work, which was started as *Essai sur les langues du Nouveau Continent* in 1812.[28] When Wilhelm failed to deliver, Alexander felt competent enough as a linguist[29] and took on the task himself. Both were in the closest exchange with each other; and for Alexander, too, the subject of indigenous languages in particular was of the utmost importance, so that it could not be omitted from his American Travels.

From Wilhelm's *Essai sur les langues du Nouveau Continent* for Alexander, a direct, albeit tortuous path led to Wilhelm's main work in the philosophy of language, his famous introduction to the *Kawi work* entitled *Über die Verschiedenheit des menschlichen Sprachbaues* (1836). Once again, differences, dissimilarity, and diversity, for all their search for unity, are at the forefront of the scholarly work. Wilhelm's studies of indigenous American languages undoubtedly had a significant

---

[26] Trabant, Jürgen: *Der Himmel, das Haus, das Gold, der gute Mann und das Nichts. Die amerikanischen Sprachen und das Weltbewusstsein der anderen Moderne.* In: Buschmann, Albrecht/ Drews, Julian/Kraft, Tobias/Kraume, Anne/Messling, Markus/Müller, Gesine (eds.): *Literatur leben. Festschrift for Ottmar Ette.* Madrid/Frankfurt a. M.: Iberoamericana—Vervuert 2016, pp. 267–280, here p. 271.

[27] Ibid., p. 272.

[28] Ibid.

[29] Cf. the article 'Sprachwissenschaft' by Jürgen Trabant. In: Ette, Ottmar: *Alexander von Humboldt Handbuch.* Stuttgart: J.B. Metzler 2018. Sowie Borsche, Tilman: Die Sprache als Medium (des menschlichen In-der-Welt-Seins). In: Messling, Markus/Tintemann, Ute (eds.): *Der Mensch ist nur Mensch durch Sprache. Zur Sprachlichkeit des Menschen.* München: W. Fink 2009, p. 69–77.

"share in the emergence of a linguistic science of human thought and thus in the development of the world consciousness of modernity"[30] in general. It is obvious that these developments, summarized here very briefly, are an important part of the realization of that scientific program that Wilhelm had formulated in 1793 in close consultation with Alexander.

Many concepts of the two Humboldts refer to each other or are at least closely interrelated. Thus, for example, Alexander's concept of the 'painting of nature' could very well be related to Wilhelm's notion of the 'total impression' (of languages, for example), since the 'painting of nature' unfolds a notion of quasi-simultaneous synopsis of a large number of different phenomena inherited from pasigraphy. It is committed to the attempt to make a high complexity of research insights and results accessible as an impression at *a glance*, as it were, and to present them to the senses.

Since Alexander, for example, in his *Tableau physique des Andes et des pays voisins*, emphasized the aesthetic dimensions of knowledge in the artistic realization of his 'nature painting' no less than the results of his research in the fields of agronomy or anthropology, geography or geology, ecology or economics, of plant geography or volcanology, one can speak here of a highly elaborated dimension of transdisciplinarity as well as of transmediality, which distinguishes Humboldt's science at the same time in the interplay of word and image, of literary and scientific expressiveness as a science that depends on the total impression. Here it becomes apparent at a glance how everything is interaction.

In Wilhelm's case, as in Alexander's, it is not a question of achieving a comprehensive but static picture, but rather the gradual development of a dynamic, as it corresponds to the conceptualizations of both brothers. Just as in Alexander's 'Naturgemälde der Tropenländer' *everything*—from the geology of drifting continents to the migrations of plants and the changing snow boundaries and climatological conditions to volcanology and zoology—is in motion, so in Wilhelm's science the principle of *energeia* proves to be the fundamental driving force of linguistic processes, but also in constitutional law and the state constitution of communities. Wilhelm is always concerned with activity, with a force that dynamically drives and changes, just as Alexander was concerned with the dynamics and vectoriality of all knowledge and thus with *knowledge from movement*.

Thus, at times a closeness, at times a commonality of the concepts of both brothers can be observed, which points to a commonality of comprehension. Thus, in this area, too, basic lines of a Humboldtian Science emerge that may apply not only to Alexander, but to both brothers.

[30] Trabant, Jürgen: *Der Himmel, das Haus, das Gold, der gute Mann und das Nichts. Die amerikanischen Sprachen und das Weltbewusstsein der anderen Moderne*, p. 273.

On Globalism: Prussia and the World

For the Humboldts, the idea of unity is based precisely not on uniformity or homogeneity, but on diversity and differences that are in constant interaction with one another. In order to be able to grasp these interactions, a science is almost inevitably required that strives to grasp the entire planet and thus to achieve globality and totality.

Alexander grasped this globality in the *summa* of his *cosmos* from a multiple relationality, in which the earth is placed in an interplay of cosmic proportions: "heaven and earth, all created things"[31] were to be grasped in the process. Such a research program was certainly not modest, since it included not least a study of the earth from the perspective of the most diverse disciplines, the most diverse logics of knowledge, including those of the arts. It was not for nothing that Alexander became the first globalization theorist on earth on the basis of long decades of historiographical study, whereby his insights not only into the acceleration phases, but also into the abortions of accelerated globalization are of enormous importance for us today—after the end of the fourth phase of accelerated globalization.

In Wilhelm's project to get to the bottom of the 'diversity of human language structure', the global dimension could not be missing either. For only the *globality* of a comparative view could guarantee that one was able to say something really general and, if possible, even generally valid about the structure and dynamics of human languages. Alexander's planetary survey of geomagnetism or the design of its globally proceeding isotherms are thus related to Wilhelm's linguistic-philosophical and linguistic-analytical investigations in a way that is both principled and functional. Humboldtian science is a world science that builds on a constantly growing world consciousness.

The stylization of Alexander as a 'natural scientist' not only belonged to a highly convenient, but certainly untenable enterprise that attempted to divide the sciences. Humboldtian Science, however, shows—regardless of whether we refer to Alexander alone or also, as happened here, include Wilhelm—that there is a unity of the sciences as well as of science that was striven for and practiced here.

If one wants to speak of a certain complementarity of Wilhelm's and Alexander's scientific activities, then it is certainly not the one that is used to distinguish between the 'Natural Sciences' on the one hand and the 'Humanities' on the other. It is not a matter of the complementarity of two different fields that are supposedly alien to each other, but rather of the redemption of the basic axiom of Humboldtian Science, namely that everything is based on interaction—including research into the interactions between man and the living world, "how he affects everything apart from him and how everything apart from him affects him".[32]

---

[31] *Briefe von Alexander von Humboldt an Varnhagen von Ense aus den Jahren 1827 bis 1858.* In addition to excerpts from Varnhagen's diaries and letters from Varnhagen and others to Humboldt. [Edited by Ludmilla Assing.] Leipzig: F.A. Brockhaus 1860, p. 22.

[32] Humboldt, Wilhelm von: *Briefe an Karl Gustav von Brinkmann*, p. 141.

The complementarity of the two Humboldt brothers—if we want to speak of a complementarity at all—is located, in terms of their conceptions, around that complex intersection of the sciences and, even more so, of the organization of science, which is misused as a dividing plane in the thesis of the *Two Cultures*. Here, too, Humboldtian Science proves to be a transnational *energeia*, a science of movement.

All this also applies to other fields of the two Humboldts' activities. It cannot be a question of understanding the one as the Prussian statesman and diplomat and thus as the 'real', the 'true' Prussia, while the other is labelled a 'world traveller' and 'cosmopolitan' and at the same time is excluded from Prussia and—as can often be observed today—expatriated. Through his numerous journeys, for example to the Iberian region, as well as his activities abroad in Paris, Vienna, Rome[33] or London, among other places, Wilhelm got to know large parts of Europe, which gave him an outstanding insight into the constitution and even more into the future possibilities of his own country. Alexander, in turn, in addition to his numerous inner-European journeys, which took him to the North and South, West and East of the continent, undertook two major transcontinental journeys, which allowed him to view Europe not only from a domestic perspective, but also from a non-European one.

For both brothers it may be said that they, as *mobile* Prussians, understood Prussia as a *mobile* and understood far better than most of their contemporaries to what extent, on a collective as well as on an individual level, the multiplicity of origins—they did not come from the traditional lines of a Huguenot family for nothing—fundamentally expands the multiplicity of futures. Was one 'more Prussian' than the other?

The absurdity of the question makes any answer superfluous. It has long been time to understand Prussia not so much as a territory and territorial state with clearly definable borders, but rather as the result of movements, migrations, and dynamics that created this space as a *space of movement* in the first place. For every space—be it that of cultures, languages or states—is created by the movements that cross it and thus vectorially generate it.

Without a doubt, Wilhelm was Prussia's leading educational and scientific politician of his time. He knew the power of knowledge. As the mastermind, head and co-founder of the Berlin University, he completed that work which he had begun with his preoccupation with teaching at the most diverse levels. That the model of 'his' university became a model not only for the German-speaking world, but also for Europe and beyond, is as undisputed as it is indisputable. It stood and stands for the unity and freedom of research and teaching.

At the present time, when the Humboldtian University is in an advanced state of disintegration and destruction, one may well ask whether it would not be high time to develop a Humboldtian University of *both* brothers, since Alexander, too, had intensively dealt with questions of education, founded a school for the children of

[33] On the intensive collaboration between the two brothers in Rome in 1805, especially with regard to the study of American languages, see Bourguet, Marie-Noëlle: *Le monde dans un carnet. Alexander von Humboldt en Italie (1805)*. Paris: Editions du Félin 2017, among others p. 145 ff and 222 ff.

miners at his own expense, constantly investigated the educational dimensions of the countries he had travelled to, and opened up problem areas of knowledge and science, not only in his *cosmos*, which are of the greatest relevance for our time. To Wilhelm's Humboldt University must be added today Alexander's Humboldt University, a university that, in the sign of its transdisciplinary and at the same time transcultural orientation and its transcending of the most diverse areas, opens itself to a world that is aware of its own history of movement and seeks to live in peace and difference, a university from the *critical conviviality* of languages.

There is no question that Alexander's Humboldtian Science is a *Life Science*—not only in the restricted sense of the medical-technological ensemble of Life Sciences, but also bearing in mind that *bios* always includes the cultural dimension of life. In the most diverse disciplines, from Anthropology to Zoology, from Biology to Chemistry, from History to Ancient American Studies, Alexander was always on the trail of life.

But the questions about the norms and possible forms of life as well as living together were not Alexander's questions alone. Wilhelm's entire life, especially his so diverse and self-sacrificing activities in the civil service, were marked by the question of how we want to live together and how we can live together. Conviviality is consequently a key concept for understanding Humboldtian scholarship both inside and outside the Humboldtian University.

Both brothers were united by the firm conviction that political decisions can only be discussed and made responsibly on the basis of sound scientific knowledge. It is important to remember this at a time when it seems to have become fatally possible to invalidate knowledge that has been persistently collected and scientifically secured over decades—such as in climate impact research—with a few populist sentences. The sometimes very intelligent fabrication of stupidity shows that the principles of an ethical foundation of science have lost none of their importance and urgency. There is no doubt that the voices of the Humboldts give weight to the function and importance of science and of scientifically based political action.

Wilhelm not only shaped the modern structure of the university, which would have to be supplemented by the world-conscious structuring of Alexander's concepts, but also the conception of a museum landscape, for whose first flagship he wanted Alexander to be the founding director. When Wilhelm's urgent request reached him, he was beyond the borders of Europe on his second transcontinental journey, his Russian-Siberian expedition. His reply from the depths of the Tsarist Empire came—as almost always in his correspondence with Wilhelm—in French and as if shot from a pistol: He had not left Paris to become director of a picture gallery in Berlin. He would rather turn his back on Prussia again than be pressed into a permanent job as commission chairman, director or president.[34]

---

[34] Cf. Ette, Ottmar: Die Aktualität Alexander von Humboldts. Perspektiven eines Vordenkers für das 21. Jahrhundert. In: Valentin, Jean-Marie (ed.): *Alexander von Humboldt. 150e annoversaire de sa mort.* Sorbonne 23–24 octobre 2009. Sondernummer der *Etudes Germaniques* (Paris) LXVI, 1 (janvier–mars 2011), p. 123–138, hier p. 124 und 129.

Unlike Wilhelm, who in his Caroline had at his side one of the most impressive female personalities of the nineteenth century alongside Rahel Varnhagen, Alexander was a convinced nomad. Public offices and obligations were not allowed to stand in the way of his desire for freedom.

Museums were by no means foreign to Alexander. He had studied their forms, contents and functions just as much in Paris or Rome as in Mexico. Like Wilhelm, the scientific landscape of Berlin was also a matter close to his heart. Let us not forget that it was Alexander who sought to realize Wilhelm's lofty ideas and plans bit by bit, particularly with regard to Berlin as a centre of science, and who succeeded—as Eberhard Knobloch has convincingly shown[35]—in equipping the Prussian capital with world-leading scientific institutions and scholars, since Berlin was to[36] receive in future "the first observatory, the first chemical institute, the first botanical garden, the first school of transcendental mathematics". And let us not forget that after Wilhelm's death in 1835, Alexander actively intervened in the design of Berlin's museum landscape from a changed perspective oriented towards his transareal world consciousness, and oriented the Museum Island towards capturing a totality of world cultures.[37] With regard to his brother, Alexander's orientation seems to me neither contrary nor complementary in a negative sense, but rather—despite all existing differences—congenial in the awareness of its fundamental effect on the 'human race'. Wilhelm and Alexander have—both together and separately—profoundly shaped Berlin's scientific and museum landscape.

Certainly, the differences between the approaches of the two brothers with regard to a very different accentuation within their Humboldtian Science could be clearly worked out. The considerations presented here were not about constructing a homogeneous, contradiction-free unity on the basis of a common origin, a brotherly shared genealogy. Beyond this genealogical connection, but also beyond a mythologizing talk of the Dioscuri, which has never contributed anything to the understanding of the pair of brothers, it is crucial to comprehensibly elaborate the fundamental

[35] Cf. e.g. Knobloch, Eberhard: 'Es wäre unmöglich nur ein halbes Jahr so zu leben wie er': Encke, Humboldt und was wir schon immer über die neue Berliner Sternwarte wissen wollten. In: Hamel, Jürgen, Knobloch, Eberhard/Pieper, Herbert (eds.): *Alexander von Humboldt in Berlin. Sein Einfluß auf die Entwicklung der Wissenschaften. Beiträge zu einem Symposium.* Augsburg: ERV 2003, p. 27–57. Vgl. auch Biermann, Kurt-R./Schwarz, Ingo: 'Moralische Sandwüste und blühende Kartoffelfelder'—Humboldt, ein Weltbürger in Berlin. In: *Alexander von Humboldt—Netzwerke des Wissens.* Catalog of the exhibition at the House of World Cultures (Berlin) from June 6 to August 15, 1999 and at the Art and Exhibition Hall of the Federal Republic of Germany (Bonn) from September 15, 1999 to January 9, 2000. Bonn: Kunst- und Ausstellungshalle der Bundesrepublik Deutschland 1999, p. 183–200.

[36] This is a letter from Humboldt to Spiker dated before 12 April 1829; cf. Humboldt, Alexander von: *Briefwechsel mit Samuel Heinrich Spiker.* Edited by Ingo Schwarz. Berlin: Akademie-Verlag 2007, 62 f.

[37] Cf. Bredekamp, Horst: Der lange Atem der Kunstkammer: das Neue Museum als Avantgarde der Vorvergangenheit. In: Bergvelt, Elinoor/Wezel, Elsa van (eds.): *Museale Spezialisierung und Nationalisierung ab 1830. Das Neue Museum in Berlin im internationalen Kontext.* Berlin: G + H 2011, p. 25–36, hier p. 33.

epistemic overlaps and even more the epistemological challenges. Alexander von Humboldt's 250th birthday is a welcome occasion for this.

For Wilhelm and Alexander von Humboldt are above all this: Challenge, obligation and impulse to translate the brothers' work into the twenty-first century. The pair of brothers was not a pair of opposites: The nomadic knowledge of the younger was not opposed by a monadic knowledge of the elder. Neither nature nor spirit separated them. The task of thinking two brothers together in a Humboldtian science will only succeed if we interrogate Wilhelm and Alexander's writings in their historical contextualization in a radical way that traces their roots. And if we endeavor to regard both brothers—united, as it were, in the salon of Rahel Levin Varnhagen—as important sources of inspiration for the future.

When Goethe, in his conversations with Eckermann on January 31, 1827, polemically called against the emerging concept of national literature, "the epoch of world literature is at hand",[38] he did not mean a supra-temporal concept, but rather an epoch that has its beginning and also its end. A single world literature has long since been replaced by the literatures of the world, which in their fundamental multilingualism and translatability, like no other discursive form, not only preserve the knowledge of mankind across the millennia, across the cultures, across the languages, but also constantly develop it further and translate it into the most diverse languages in the long term.

The multi-logical structures of the languages of the world as well as the Literatures of the World show us the possible forms and norms of a life in different logics at the same time. The coexistence between people of different origins, but also with the gods, the animals, the plants as well as the whole of nature, as we know it since the *Gilgamesh epic* or the *Shi-Jing*, points the way to a conviviality that should be at the heart of a Humboldtian forum.

Beyond a thinking in the dialogical, which all too easily slides into an alterization, an *othering* of the other, who is then inferiorized and fought against, it is a matter of thinking further, of an epistemology of expansion, which does not dissolve into ever new front lines. Today, more urgently than ever, we need forums of the multi-logical: schools and laboratories in which we learn to think in different logics at the same time.

Humboldtian Science is a Life Science in the fullest sense: it asks about life from the diversity of cultures, languages and ways of life. It asks about conviviality, about the possibilities and limits, opportunities and risks of living together in a community, a country, a union of countries, and on a global scale. We could call this an *ecology of the polylogical.* Above all, however, we could learn from a formula that has been played out here, which puts nothing first and which does without any hierarchization: *everything is interaction.*

[38] Eckermann, Johann Peter: *Gespräche mit Goethe in den letzten Jahren seines Lebens.* Edited by Fritz Bergemann. Bd. I. Frankfurt a. M.: Insel 1981, p. 211.

From Georg Forster to Adelbert von Chamisso: A Voyage Round the World

Georg Forster (1754–1794)

World History and Vectopia

In 1774, no less a figure than Johann Gottfried Herder, in his writing *Auch eine Philosophie der Geschichte zur Bildung der Menschheit*, very pointedly described that tremendous movement which we may call the second phase of accelerated globalization[1] and which triggered a dynamic spreading from Europe across the entire world, to which philosophy, even in the German-speaking world, which was marginal in terms of world politics, was by no means alien. For it was seized by the force of a dynamic that quickly—decades before Hegel—propelled it into world-historical questions. Did not the sciences and many scientists see themselves caught up in a frenzy of movement that quickly overtaxed the tableau-shaped possibility of arranging knowledge and was to bring about the end of natural history[2] at the end of the eighteenth century?

This constantly accelerating dynamic in the second half of the *Siècle des Lumières*—and here the seventies and eighties formed a clear climax—this vectorization of all things and all senses undoubtedly constituted the decisive feature of the epoch. Consequently, Johann Gottfried Herder could refer with good reason and with moving, sometimes agitated words precisely to the leading genre of travelogues of the time and the wealth of material brought to light in them, not without a twinkle in his eye:

> Our travelogues multiply and improve; everything runs, which has nothing to do in Europe, with a sort of philosophical fury over the earth-we collect 'materials from all the world's

[1] Cf. on this Ette, Ottmar: *TransArea. Eine literarische Globalisierungsgeschichte.* Berlin/Boston: de Gruyter 2012.

[2] Cf. Lepenies, Wolf: *Das Ende der Naturgeschichte. Wandel kultureller Selbstverständlichkeiten in den Wissenschaften des 18. und 19. Jahrhunderts.* Frankfurt a. M.: Suhrkamp 1978.

O. Ette, *Mobile Prussia*, https://doi.org/10.1007/978-3-476-05839-3_8

> end,' and shall once find in them what we least sought, discussions of the history of the most important human world.[3]

The young Georg Forster was undoubtedly one of those highly vectored protagonists who, in this history of the expansion of knowledge—and not without 'philosophical rage'—was able to put the most expressive material into circulation. But he was by no means uncritical of the developments ironically brought to the point by Herder.

Rather, he tried to develop fundamental insights into the epistemology of knowledge about the non-European world in his literary-philosophical work, in order to expand his experience of the world—and not just "the most important human world"[4]—in such a way that a new, fundamentally expanded understanding and experience of the world could emerge from it. Not for nothing did the young Alexander von Humboldt follow in his footsteps; and not for nothing did Adelbert von Chamisso still pay him a visible homage in the title formulations of his travel work.

In his 'Preface' to his *A Voyage Round the World*, dated London, March 24, 1777, the young Georg Forster presented the epistemological positioning of his travelogue and thus at the same time recorded the key data for a paradigm shift in travel theory and travel literature at the end of the eighteenth century:

> The philosophers of this century, who were very displeased with the apparent contradictions of various travelers, chose certain writers whom they preferred to the others, giving them all credence, while regarding all others as fabulous. Without sufficient knowledge, they set themselves up as judges, assumed certain propositions to be true, (which, moreover, they distorted according to their own discretion,) and in this way built systems for themselves, which, from a distance, are obvious, but, on closer examination, deceive us like a dream with false appearances. At last the scholars grew tired of being carried away by declamation and sophistical reasons, and loudly demanded that only facts should be collected. Their wish was granted; facts were collected in all parts of the world, and in all this their science was no better off. They got a mingled heap of loose individual links, from which no art could produce a whole; and by hunting for facts to the point of nonsense, they lost all other attention, and became incapable of determining and abstracting even a single proposition; like those micrologists who spend their whole lives on the anatomy of a mosquito, from which, however, not the slightest consequence can be drawn for men and cattle.[5]

With his characteristic critical clarity, Georg Forster basically opposes two opposing positions in the history of science, not only in the eighteenth century. On the one hand, he vehemently rejects the works of those *philosophes* who—such as the Dutchman Cornelius de Pauw in his *Recherches philosophiques sur les Américains* or the Frenchman Guillaume-Thomas Raynal in his *Histoire des deux Indes*—rose to judge the truthfulness of travelogues and other texts, without ever having set foot on non-European soil and undertaken journeys themselves, and without being able

[3] Herder, Johann Gottfried: *Auch eine Philosophie der Geschichte zur Bildung der Menschheit.* Frankfurt a. M.: Suhrkamp 1967, p. 89.

[4] Ibid.

[5] Forster, Georg: *Reise um die Welt*, p. 16 f.

to judge the writings they considered true or false from an empirically founded knowledge on the ground. They lacked a sensual, physical, close-up experience of a world that they—as philologists *avant la lettre*,[6] as it were—believed to know very well just by reading texts.

The Berlin debate about the New World, which began with the publication of the first volume of the *Recherches philosophiques* in 1768 and reached its first climax in the sharp exchange of blows on September 7, 1769, with the counter-speech of Antoine-Joseph Pernetty before the Berlin Academy, had brought this fundamental problem of the *armchair travelers* to the attention of a readership outside Europe as well.[7]

Georg Forster was at the height of the debates of his turbulent time and knew beyond doubt that especially the New World readers of the two European *philosophes* had vehemently denounced this ignorance of the conditions on the ground and demanded an empirical familiarity also and especially of the European scholars with their non-European objects. A knowledge of the world could no longer be based on philological textual knowledge alone—even if the *géographes de cabinet*, like the best cartographers of their time such as a D'Anville who never went further than forty miles from Paris, continued to map the surface of the earth without ever having seen it.[8] But this, if we follow Georg Forster, was only half the problem.

For, on the other hand, Forster also distanced himself from an empirically oriented collection of facts that proceeded without sense and understanding and, as it were, got lost 'micrologically' in the most unimportant details, engaged in—as Forster suggests—fly-leg counting and thereby lost sight of that totality: that world which, in its inner connections, was so dear to the author of the *A Voyage Round the World*.

It is not without reason that the term 'world' is always understood by Georg Forster in its etymologically 'human' meaning as a totality that can only be grasped as a whole. His experience of the world, as expressed in his travel literature, was aimed at experiencing this totality.

On and after his circumnavigation of the world with James Cook he himself was by no means concerned with stringing together as many isolated 'facts' as possible, but as far as possible with placing the observations made on the spot in larger contexts, without thereby falling into those chimerical 'systems' that without empirical basis pursued a textual science based only on references to other texts—a kind of philology in the negative sense. Georg Forster was neither concerned with systems

[6] Cf. on this Ette, Ottmar: Wörter—Mächte—Stämme. Cornelius de Pauw und der Disput um eine neue Welt. In: Messling, Markus/Ette, Ottmar (eds.): *Wort Macht Stamm. Rassismus und Determinismus in der Philologie (18./19. Jh.)*. With the collaboration of Philipp Krämer and Markus A. Lenz. Munich: W. Fink 2013, pp. 107–135.

[7] Cf. on the Berlin debate the Potsdam conference volume by Bernaschina, Vicente/Kraft, Tobias/Kraume, Anne (eds.): *Globalisierung in Zeiten der Aufklärung. Texte und Kontexte zur 'Berliner Debatte' um die Neue Welt (17./18. Jh.)*. 2 parts. Frankfurt a. M./Bern/New York: Peter Lang Edition 2015. As well as the second and especially the third chapter in the present volume.

[8] Cf. Broc, Numa: *La Géographie des Philosophes. Géographes et voyageurs français au XVIIIe siècle*. Paris: Editions Ophrys 1975, p. 34.

nor with fly-legs: in his research and in his thinking—like Alexander von Humboldt after him and in his footsteps—he was simply concerned with the whole.[9] This is precisely where the deep kinship between the two German travelers lies.

As is well known, Georg Forster, together with his father Reinhold, had accompanied James Cook on his second voyage around the world and, since unlike his father he was not contractually forbidden to do so, had become the author of the report of his *Voyage Round the World*,[10] which was published in 1777, first in English and then, from 1778 to 1780, in a German edition—and which was so highly regarded by Alexander von Humboldt. The enormous resonance of this travelogue, which was as extensive as it was aesthetically sophisticated, transformed Georg Forster into the emblematic figure of the world traveller par excellence in Germany almost overnight.

And Forster remained on the cutting edge, even on a global scale. In his writing on 'The Northwest Coast of America', published in 1791—i.e. after the French Revolution—he took the seemingly marginal trade in the north of the American continent as an opportunity to engage in more fundamental considerations, which in a sense already took stock of the second phase of accelerated globalization:

> The time is approaching with rapid steps when the whole earth will be revealed to the European spirit of investigation, and every gap in our empirical sciences will have to be filled, if not entirely, yet to such an extent that we shall be able to survey the connection of things more fully, at least at the point in the ether which we inhabit.[11]

With all the "gaps" filled in, the scope for those utopias that emerged under the sign of 'Thomas Morus' *Utopia* from 1516 in the context of the first phase of accelerated globalization has long since become considerably smaller and increasingly diminished. Forster's remarks incidentally draw attention to the fact that the dwindling space for the *u-topos* at the end of the eighteenth century gives rise to a situation in which now not space but time becomes the projection surface of what is longed for, or at least the testing ground for what is imagined: The u-chronic[12] emerges and

[9] Cf. also the essay by Schmitter, Peter: Zur Wissenschaftskonzeption Georg Forsters und dessen biographischen Bezügen zu den Brüdern Humboldt. A preliminary study on the relationship between 'general natural history', 'physical description of the world' and 'general linguistics'. In: Naumann, Bernd/Plank, Frans/Hofbauer, Gottfried/Hooykaas, Reijer (eds.): *Language and Earth: Elective Affinities between the emerging Sciences of Linguistics and Geology.* Amsterdam: Benjamins 1992, pp. 91–124.

[10] Cf. Steiner, Gerhard: Georg Forsters 'Reise um die Welt'. In: Forster, Georg: *Reise um die Welt.* Edited and commented with a afterword by Gerhard Steiner. Frankfurt a. M.: Insel 1983, p. 1015.

[11] Forster, Georg: Die Nordwestküste von Amerika, und der dortige Pelzhandel. In Georg Forster: *Works. Complete writings, diaries, letters. Sämtliche Schriften, Tagebücher, Briefe.* Bd. V: Kleine Schriften zur Völker- und Länderkunde. Edited by Horst Fiedler, Klaus-Georg Popp, Annerose Schneider und Christian Suckow. Berlin: Akademie-Verlag 1985, p. 390.

[12] Cf. Krauß, Henning: Der Ursprung des geschichtlichen Weltbildes, die Herausbildung der 'opinion publique' und die literarischen Uchronien. In: *Romanistische Zeitschrift für Literaturgeschichte* (Heidelberg) XI, 3–4 (1987), pp. 337–352; and bei: Krauß, Henning: La 'Querelle des Anciens et des Modernes' et le début de l'uchronie littéraire. In: Hudde, Hinrich/Kuon, Peter (eds.): *De l'utopie à l'*uchronie. Actes du colloque d'Erlangen, 16–18 octobre 1986. Tübingen: Narr 1988, p. 89–98.

finds its epoch-making literary expression in Louis-Sébastien Mercier's *L'an 2440*,[13] which continues to have an impact on George Orwell's past future worlds.

At the same time, however, in the passage just cited Georg Forster links two isotopias that are alluded to in the title of the present reflections. On the one hand he refers to an increasingly complete knowledge of the world in the light of the empirical sciences, brought about by the European travelers, to which Herder had already alluded; and on the other hand to the fact that thereby also the life at a certain "point in the ether" changed by the growing insights into the "connection of things"—and this certainly not only on a level of scientifically verifiable knowledge.

The reference to the changed state of knowledge, especially of European science, as the result of that process of expansion that had contributed to the end of the natural-historical tableau through the increased circulation and accumulation of the most diverse fields of knowledge, not only changes scientific epistemologies, but also the life and experience of a world that has become considerably wider for contemporaries. Not only had "our present physical and statistical knowledge of Europe reached perfection, but also the most distant parts of the world" had emerged from "the shadows in which they lay buried only a short time ago".[14] To use a Humboldtian term: world consciousness[15] had long since begun to change in the context of the second phase of accelerated globalization.

The metaphor of light of the Enlightenment throws a clear light on Forster's own point of view: James Cook's voyages to the northwest coast of America would have brought many new discoveries, so that "without him, the fur trade between China and this newly discovered coast would hardly have come about, and a collision would have arisen between the courts of Madrid and London".[16]

The Russian-American Trade Company, founded as recently as 1799, without whose existence Adelbert von Chamisso's voyage around the world would not have taken place, had not yet intervened in these circumstances. Since the end of the Seven Years' War, however, there had been a growing awareness throughout Europe that local disputes could quickly turn into regional, even global conflicts, which could lead to, or at least contribute to, fundamental upheavals, especially outside Europe—as the long-lasting process of *Independencia* was to show only a few decades later in the American hemisphere. The problem of global conviviality had long since ceased to be an abstract philosophical question: it had become an everyday fact.

---

[13] Cf. Jurt, Joseph: Louis-Sébastien Mercier et le problème de l'esclavage et des colonies. In: *Anales del Caribe* (Santiago de Cuba) 7–8 (1987/1988), pp. 94–107; as well as: Stadtreform und utopischer Entwurf: von Alberti bis L.-S. Mercier. In: Hahn, Kurt/Hausmann, Matthias (eds.): *Visionen des Urbanen. (Anti-)utopische Stadtentwürfe in der französischen Wort- und Bildkunst.* Heidelberg: Winter 2012, pp. 21–31.

[14] Forster, Georg: *Die Nordwestküste von Amerika, und der dortige Pelzhandel*, p. 393.

[15] Cf. Ette, Ottmar: *Weltbewußtsein. Alexander von Humboldt und das unvollendete Projekt einer anderen Moderne.* Weilerswist: Velbrück Wissenschaft 2002.

[16] Forster, Georg: *Die Nordwestküste von Amerika, und der dortige Pelzhandel*, p. 395.

Precisely for this reason, it remained a burning philosophical challenge. Certainly, warlike conflicts triggered by such worldwide clashes of interests are not in the foreground of Forster's almost prophetic world-historical observations. Rather, the enormous acceleration he observed led Forster to conclude that a new historical epoch had begun that would undoubtedly be global in character:

> Here begins a new epoch in the strange history of the European trade, this trade, in which the whole world history seems to gradually dissolve. Here, so many ideas and facts impose themselves on the researcher that it seems to be the duty of the publisher of the new voyages and land voyages in that region to collect everything that relates to the knowledge of the same in a focal point and, especially for an audience like ours, which can only take a literary, indirect interest in the discoveries of the sea powers, to facilitate the overview of what has been done so far and the judgment about the importance of this whole matter.[17]

This passage is so revealing for us not only because it reflects the specific situation of a German-speaking readership which—not belonging to the audience of the maritime and colonial powers—had no direct colonial interests, but was 'only' oriented towards a literarily mediated increase in knowledge. Prussia, too, had long since ceased to participate in the transatlantic slave trade in a colonialist-exploitative manner by this time. This dimension distant from colonialism must be taken into account in the further course of our investigation in German-language publications, although not (as in the case of Forster or Humboldt) in English- or French-language publications.

Above all, however, Georg Forster's thesis, based on the fur trade in the North of America,[18] of a new epoch in which the whole of world history would evaporate into world trade, is of such great significance for the philosophy of history that one must emphasize both the far-sightedness of a subtle reflection on the consequences of this second phase of accelerated globalization, in which the author of *A Voyage Round the World* himself was involved, and the more general phenomenon that phases of acceleration produce a stroboscopic effect, as it were, in the philosophy of history. The experience of history changes: in phases of rapid acceleration it is as if a history begins after history.

For just as the wheels of carriages and covered wagons in the classic Western seem to stop precisely when they reach their highest speed, it is precisely in the phases of historical acceleration that the impression of immobilization, of an *after-history*, seems to arise, the perception of which always seems to run parallel to the emergence of modern historical thought.[19] Paradoxically, the impression of highest rest is most easily created in the center of the vortex of accelerated movement.

But one could also venture the thesis that it is precisely phases of the disintegration of an accelerated globalization—as we can clearly observe at the present

---

[17] Ibid.

[18] Here Forster drew, among other things, on a writing by Alexander Dalrymple on the fur trade; cf. ibid., p. 778.

[19] Cf. on this Ette, Ottmar: *Literatur in Bewegung. Raum und Dynamik grenzüberschreitenden Schreibens in Europa und Amerika.* Weilerswist: Velbrück Wissenschaft 2001, pp. 9 and 539.

time—that are experienced as a rapid multiplication of the most diverse local and regional conflict developments, which are admittedly still capable of bringing about worldwide changes. Unlike Georg Forster, we find ourselves today, at the end of the fourth phase of accelerated globalization, in a situation in which a history after history, a 'Posthistoire Now!' as it were, can no longer be experienced.[20]

The deceleration of globalization, as it currently confronts us in a wave of new nationalisms, protectionisms and fundamentalisms, is therefore not to be equated with an experience of political, historical or social calm.

The 'way out' of the projection that jumps from utopia to uchrony due to the lack of unknown spaces is certainly not the only way to 'cope' with a space-time problem that arose as a consequence of Western expansion and scientific experience of the world. For Georg Forster as well as Alexander von Humboldt or Adelbert von Chamisso chose ways of reacting that were related to their own lives and thus *livable*, and that did not refer to space alone or to time alone, but to a specific connection of the dimensions of space and time. They developed and *lived* what at this point should be introduced conceptually as an—admittedly gräcolatinized—neologism: the *vectopia*.

The concept of vectopia stands for a linking of the projection surfaces of utopia and uchronia in space and time in a way that makes the kinetic[21] dimension, the experience and the experiencing of movement and even more so of vectoricity, the actual means of trying out and experiencing the world. This globally extended experience of the world at the same time involves a further-living, which at this point in our argumentation is to be understood first of all in a spatial and not yet temporal sense and consequently primarily concerns the radius of what Forster calls 'experiential knowledge'.

Unlike utopia and uchronia, vectopia involves a material dimension related to the body and the corporeal, and thus a life and experience of the world that cannot do without the constant change of place, without a travel that is taken up again and again. Vectopia unfolds the projections of a life not from space, not from time alone, but, thanks to their combinatorics, from a *vectoricity* in which all previous movements are stored and all subsequent movements are already laid out. Vectopia is more than a figure of thought: it is vitally linked to life and thus a figure of life.

Vectopia and Continued Life

Georg Forster demonstrated that the experience of the world-wide, of that which encompasses the whole world, is not bound to a 'voyage round the world' of any

[20] Cf. Gumbrecht, Hans Ulrich: Posthistoire now. In Gumbrecht, Hans Ulrich/Link-Heer, Ursula (eds.): *Epochenschwellen und Epochenstrukturen in der Literatur- und Sprachhistorie.* Frankfurt a. M.: Suhrkamp 1985, pp. 34–50.

[21] The Greek compound of a 'kinetopia' would have the disadvantage of being too strongly confronted with predominantly cinematographic, but also kinesiological connotations and would therefore lead to misunderstandings.

kind, not by the example of a utopia, a heterotopia or an atopia, but by a vectoropia in the literary midst of Western Europe:

> In one night our scene has changed so much that nothing that exists at present awakens a trace of yesterday in our memory. We live in a different world, with people of a different kind.[22]

But how is this different experience of a different world from yesterday to today to be understood? Georg Forster begins the twenty-fifth chapter of his *Ansichten vom Niederrhein, von Brabant, Flandern, Holland, England und Frankreich im April, Mai und Junius 1790* with the words just quoted; and in these words their author presents the journey between two places as a journey into a new, into another world. But what is this other world like?

This is by no means a literary whim of the well-travelled author. Rather, the movement of the travelers from The Hague to Amsterdam proves to be a traveling movement that connects places and people *in different worlds* with each other through words, as it were in transit, and at the same time contrasts them reciprocally. In doing so, the discontinuity so masterfully staged here by Georg Forster opens onto that complexity of movement that has been such a literary aesthetic feature of his literary writing in his travelogues since his *Voyage Round the World.* For the world traveler's account offers far more than views from the Lower Rhine or views from Western Europe. With the Netherlands, and in the further course of the journey also England and France, three of those 'sea powers' so apostrophized by him are visited, which are distinguished by global claims to power and worldwide accumulated wealth.

It is certainly not primarily the basic itinerary structure, the literarily unfolded movement of the travelers in the travelogue, which is responsible for the generation of vectorial complexity that Forster's writing always so impressively offers. Rather, a landscape is unfolded before the reader's eyes, which in this case presents and represents an urban space of movement, an urban landscape located in the west of continental Europe, which may be understood as an urban *vectopos.*

Before we even get a glimpse of the great buildings of the city of Amsterdam, we are taken to the massive Admiralty dockyard:

> In admirable order lay here, with the marks of each particular warship, in many chambers the anchor ropes and smaller ropes, the ship's blocks and sails, the crude ordnance with its ammunition, the shotguns, pistols, and short weapons, the lanterns, compasses, flags, in a word, everything down to the least necessities of equipment. Before us spread the immense expanse of water of the harbour, and in the dim distance flashed the sands of the shallow shore beyond. Far down to the left stood the forest of many thousands of merchantmen's masts; the sun's rays played on their shining varnish.[23]

[22] Forster, Georg: *Views of the Lower Rhine.* Munich: The Hilliard Collection—Edition Deutsche Bibliothek 1985, p. 348.

[23] Ibid., p. 348 f.

A city opens up in front of the city. A vectopic space emerges from the movement of the travelers. Are we not dealing here with a city beneath the city, with a city that consists only of movements and can only be understood as a mesh of movements? In this cleverly conceived panorama of an enormous shipyard, the first thing that comes into view is that violence which forms the basis for all that wealth, all that greatness, which is spread out and listed here with its small as well as its large weapons, from the individual pistol to the violent machinery of an entire warship, as in an arsenal, as in a warlike arsenal. This city is not made of harmless movements: here we encounter one of the centers of worldwide European colonialism.

But this close-up of a material culture, which is superimposed in all its dimensionality and violence, is transformed in a second step into a landscape that allows the gaze to glide over the "immense expanse of water of the harbor" and over the opposite shore in the distance to that "forest of many thousands of mast trees of the merchants"[24] in whose presence the elements of nature and culture, of water and harbor, of trees and masts, of wind and sails are indissolubly linked. Nature cannot be designed in any other way than from culture, from cultures. And these cultures, for their part, can only be understood in terms of the history of movement, in vectorial terms.

In this amphibious landscape rising between sea and land, between water and sand, above a forest whose trees have long since become masts and thus have moved from land to water, a panoramic view is unfolded which not by chance passes from the Admiralty's navy to the merchant fleet of the merchantmen, indeed glides gently from ship to ship. For without the firepower of the navy the economic power of the merchant fleet would be inconceivable, both forming only in their interplay those rich foundations on which the city of Amsterdam could develop in its splendor—under the protection of that "magnificent arsenal" which "resting on eighteen thousand piles and completely surrounded by water" is sunk into this landscape painting. The trees that have become poles also carry the city as subterranean forests, whose fate is tied to the mobile, globally moving element of water.

Thus, at this point in the *Ansichten vom Niederrhein*, quite in keeping with the well-calculated beginning of this chapter, another world opens up that would be inconceivable without all these movements, a world that quickly literally emerges from the water in its truly global proportions and stands before our eyes. While inspecting one of these warships, and out of the excitement it provokes at the "marvelous construction of this monstrous machine,"[25] a little farther to the right, "the ships of the East India Company," with their own shipyards, appear, extending as far as the island of Osterburg: "The arriving and departing craft, together with the little rowing boats, enlivened the scene."[26] We are placed in the middle of a vectorial landscape: This literary, but downright cinematographic, movement picture is not about spatial history Georg Forster is concerned with a history of movement.

[24] Ibid., p. 349.

[25] Ibid., p. 349.

[26] Ibid.

Behind the vehicles of the East India Company, in addition to the war machine, a trade machine now appears, which, together with the military firepower, is responsible for the immense abundance of power that Amsterdam and the Netherlands have long been able to acquire, to conquer, with their highly modern and world-spanning colonial machinery. Thus the I also logically pauses for a brief moment in its own movement and fixes this "scene of the most comprehensive activity," since "the city and even the republic owe their existence and their greatness" to this truly world-encompassing activity.[27]

Yet the traveller's view of this global power of city and state is a thoroughly positive one, since the people had once mustered a courage comparable to no other people in Europe, "to wage the Eighty Years' War with Philip the Tyrant, the powerful ruler of both Indies, and his successors".[28] Republican pride resonates here in every line of Georg Forster, the consistent supporter of the French Revolution.

The long struggle of the Dutch not only against the European occupying power, but even more against the world power of Philip II, who ruled over huge colonial territories *en las dos Indias*, in both Indies, here flashes on a few lines a history of globalization, in which the traveler Georg Forster himself was very well able to place himself, since he was allowed to see himself very well as part of this history of globalization. For he cleverly integrates Spain's early supremacy as a world power, a Spain that, together with Portugal, formed the alliance of hegemonic Iberian powers that dominated the first phase of accelerated globalization from the southern western edge of Europe.[29]

But the centuries-long history of globalization had by no means come to a standstill. Georg Forster knew himself to be one of the protagonists in another, a second phase of the acceleration of these globalization processes, which were now, however, no longer led by Spain and Portugal, but by England and France. Through his accompaniment of James Cook and his literarily polished travelogue,[30] however, he had become not only the protagonist, but even more the thinker and philosopher of this second phase of accelerated globalization, which, moreover, made him a writer of European standing who was read worldwide with the publication of his *A Voyage Round the World* in both English and German.

In his *Views of the Lower Rhine*, much of this great world journey and much of this world success resonates. On the basis of his world-spanning experience of the world, Georg Forster had a wider view than the vast majority of his contemporaries in Germany and Europe. And his constant traveling was an expression of his vectopia until the end of his short life in Paris.

But back to Forster's travel-literary movement picture of Amsterdam. In the case of the author of these *views*, we are dealing with a travel writer who is highly

[27] Ibid., p. 350.

[28] Ibid.

[29] On the various phases of accelerated globalization, cf. the opening chapter in Ette, Ottmar: *TransArea. Eine literarische Globalisierungsgeschichte* (2012).

[30] Cf. Steiner, Gerhard: Georg Forsters 'Reise um die Welt'. In: Forster, Georg: *Reise um die Welt.* Ed. and with an epilogue by Gerhard Steiner. Frankfurt a. M.: Insel 1983, p. 1015.

sensitized to such globalization thrusts and who knew very well how to assess the full significance of the military and commercial power of the Netherlands—especially since he was obviously able to classify the role of the Netherlands as a kind of 'intermediate link' between these two major phases of accelerated globalization. For even if the Netherlands, in its expansion, drew on a whole series of military and political strategies that the Iberian powers had already developed, the global expansion of this territorial state, small even in European dimensions, marked at the same time, precisely with the founding of its extremely successful East India Company, a decisive step towards a new type of world trade. New, more efficient forms of capital formation,[31] knowledge transfer and a global trading system pointed the way, which in the second half of the eighteenth century was to be established by the supremacy of the British and the French. Europe had long since ceased to be understood solely in terms of Europe—Forster's city view of Amsterdam was an impressive demonstration of this.

By those changes, which had to impose themselves on an attentive visitor of Amsterdam in 1790, Georg Forster made recognizable that and why in Amsterdam—as he wrote at the beginning of the twenty-fifth chapter—one not only lived "in a different world", but also had to deal with "people of a different kind".[32] So Forster noted right at the beginning of his twenty-sixth chapter of his *Ansichten vom Niederrhein*, also still dedicated to Amsterdam:

> In the enervating climate of India, the European conquerors become accustomed only too easily to Asiatic, soft opulence and splendor. If afterwards the restless feeling, with which they sought happiness and satisfaction there in vain, drives them back to Europe with their gold, they transplant the oriental customs into their fatherland. It is true that in republics one resisted for a time the introduction of luxury; but excessive wealth infallibly brings it in its wake. Although sober abstinence multiplied the savings of industriousness for several generations, yet at last the accumulated capital comes to a laughing heir, who, beyond the apprehension of being able only to diminish it, learns to rhyme the demands of acquisitiveness with the gratification of his senses.[33]

In this way, the great metropolises of the European colonial powers—and especially the urban vectopus of Amsterdam—were marked by a global transfer of capital, luxury goods and an 'oriental' lifestyle, which, in Forster's sense, had to lead sooner or later to a fundamental transformation of these metropolises themselves. For the long-lasting history of transfers, which in the Netherlands emanated from the East India Company in particular, gave rise to a change and transformation of those ways of life and norms of living that led the inhabitants of Amsterdam, too, to change their lifestyles, to live as if in another world and to mutate into other people. But was this also an experience of the world in the form of a life that was further thought of and further lived?

---

[31] Cf. Ferguson, Niall: *Civilization. The West and the Rest.* New York: Penguin Books 2011, p. 196 f.

[32] Forster, Georg: *Ansichten vom Niederrhein*, p. 348.

[33] Ibid., p. 359.

It remains to be said: In this vectorial sense, Amsterdam is far more than just Amsterdam: the commercial metropolis opens up to an entire world, for which the movement space of the port advances to become the actual transshipment point, a vector field of global proportions. In this way, other landscapes become visible beneath the harbour landscape of Amsterdam: other cities, for example in Asia, but also in the Caribbean, emerge beneath the cityscape of the Dutch metropolis. For they have long had a decisive influence on life in Amsterdam.

Beneath the landscape on the surface lie other landscapes, beneath the city other cities—and while each word always carries its place within it, the vectoricity unfolded in the words of the literary travelogue always allows the words beneath the words to become recognizable as places beneath the places. These are not static landscapes, but vectorial landscapes that open onto ever new landscapes on a global scale. The urban landscape of Amsterdam can only be understood as an urban vectopos.

Thus the *Ansichten vom Niederrhein* are by no means limited to the Lower Rhine, but open up in the genre and medium of travel literature to a world that the world traveller Georg Forster knew how to shape in an aesthetically convincing way as part of a global, world-spanning context and coexistence. It is a world that, from the space of movement of the port, fills everything with movement, everything with life—a life that can be captured very precisely in its acceleration:

> The city with its wharfs, docks, warehouses, and factory buildings; The bustle of the busy swarm of bees along the immense shore, on the streets and canals; The magic-like movement of so many sailing ships and boats on the Zuider Zee, and the restless swing of the thousands of windmills around me-what indescribable life, what boundlessness in this sight! Commerce and navigation embrace and use for their ends many a science; but gratefully again they offer it aid to its perfection.[34]

It is the sight of life in acceleration, of life accelerating enormously, in which "distant parts of the world are brought together" and different zones and nations are brought together.[35] It is, in this (also consumptive) sense, a life further than that in zones of lesser global compression and interconnectedness. All this takes place under the sign of a circulation of knowledge whose global dimensions no longer allow us to resort to the same words; languages too, the multilingual Forster knew, had to change. An enormous "wealth of concepts" is now being circulated "ever more rapidly",[36] a wealth of lexemes that must keep pace with the accelerated processes of differentiation not only in the realm of the sciences. The place opens onto many places, the word opens onto many words—and to represent the movement of this accelerating life, the travelogue is undoubtedly the adequate literary form of expression for Georg Forster. Was Forster not a master of this genre—and at the same time the travel-literary and travel-technical teacher of Alexander von

[34] Ibid., p. 351.

[35] Ibid.

[36] Ibid.

Humboldt, who incidentally also accompanied him on this journey to the Lower Rhine, to England and to revolutionary Paris?

The travel-literary staging of Amsterdam undoubtedly constitutes a perfect and paradigmatic example of a vectorial way of experiencing the world and living on. For in its aesthetic forms, a place like the Dutch metropolis of world trade at that time can only be understood from its global contexts. Consequently, Amsterdam is portrayed with good reason—and here the literary skill of the travel writer becomes apparent—starting from its port, since the artistic re-presentation of Amsterdam in the travelogue can only succeed through the depiction of the global vectoriality embodied by the war fleet and the merchant fleet.

For Georg Forster, experiencing the world is living on. In this sense, Amsterdam is a highly vectorized landscape, a cityscape in and out of motion, which could never be adequately grasped by a static spatial history without this widening and expansion. Thus, for Georg Forster, this urban vectopos represents a vectopia that admittedly would never know how to tie itself back to a fixed place, to a fixed residence alone.

To be sure, for Forster there was this place of political hopes, this place where he hoped to see not only the history of a nation, not only the history of a continent, but also the history of humanity in general come to completion. Revolutionary Paris, which was to mark the endpoint of his life's journey just a few years later, with the world-spanning pathos of a humanity that here proclaimed its universal claims and rights, undoubtedly became the best embodiment of that vectopia that shaped Georg Forster's life and continued to do so.

Alexander von Humboldt (1769–1859)

Vectopia and Fulfilled Life

Shortly before his death in the French capital on January 10, 1794, Georg Forster, in the winter of 1793, in the seventh letter of his *Parisische Umrisse*, the last work from his pen, stated unequivocally, with clear-eyed pride and that revolutionary spirit which distinguished him throughout his life: "Paris is always *our* map, and you have lost."[37]

This sentence from his Parisian exile, hurled at the reaction in the German lands, drew attention to the far-reaching consequences of the French Revolution, but also to that outstanding position of the *Ville-lumière*, which could have been signed not

[37] Forster, Georg: Parisische Umrisse (1794). In: Günther, Horst (ed.): *Die Französische Revolution. Berichte und Deutungen deutscher Schriftsteller und Historiker.* Frankfurt a. M.: Deutscher Klassiker Verlag 1988, p. 649. On the significance of Paris, especially in the Franco-German context, cf. Lüsebrink, Hans-Jürgen: 'La Patrie des Droits de l'Homme.' On the identification of German scholars of France with the France of the Enlightenment and the French Revolution. In: Grunewald, Michel/Lüsebrink, Hans-Jürgen/Marcowitz, Reiner/Puschner, Uwe (eds.): *France-Allemagne au XXe siècle. La production de savoir sur l'Autre.* Vol. 2: *Les spécialistes universitaires de l'Allemagne et de la France.* Bern/Frankfurt a. M./New York: Peter Lang 2012, pp. 301–314.

only by Alexander von Humboldt, who first experienced revolutionary Paris alongside Forster, but also by Adelbert von Chamisso, whose family had fled to Prussia before this very revolution. For all three world travellers, Paris, the capital of the *République des Lettres*, was probably the decisive point of reference in their lives, however important Prussian Berlin may have been for a Humboldt or Chamisso. From this point of view, the *Ansichten vom Niederrhein*, with their perspective vectopos Paris, seem like a never-arriving, never-ending journey into revolution. From revolution to revolution.

The early encounter and journey with Georg Forster can hardly be overestimated in its significance for the young Alexander von Humboldt. As a friend and teacher, the author of *Ansichten vom Niederrhein* was not only the traveling companion of the younger of the two Humboldt brothers, who was born in 1769, but also a role model and literary model for a writing that, with *Ansichten der Natur*, soon created a book that, although differently conceived, was similarly multi-perspectival in structure and, incidentally, very successful with the German-speaking public. In his 1806 French-language work entitled *Mes confessions*, which was as self-deprecating as it was tongue-in-cheek and reminiscent of the creator of modern autobiography, Jean-Jacques Rousseau, Humboldt noted:

> In the spring Mr. Georg Forster, whose acquaintance I had made at Mainz, proposed to me to accompany him to England on the rapid voyage which he has described in a little work (Ansichten etc.) justly famous for the elegance of its style. We went to Holland, England, and France. This journey cultivated my mind, and strengthened me more than ever in my resolution to travel outside Europe. For the first time I saw the sea then at Ostend, and I remember that the sight made the very greatest impression upon me. I saw not so much the sea as the lands to which that element should one day carry me.[38]

The dream, or perhaps even more the phantasm, of a (then still unspecified) non-European journey had long hovered before the young Humboldt's eyes. Since the memorable journey with Forster, however, it began to take concrete shape as a vectopia—as science and life from movement. In this text from 1806, i.e. after the completion of his great voyage to the American Tropics, which he undertook from 1799 to 1804, one can still sense the full power of fascination that had emanated from Georg Forster on his still young traveling companion. It is still noticeable in Humboldt's late writings—and let us not forget that the author of *Cosmos* not only outlived the other two world travelers by several decades, but also published over a significantly longer period of time during his life, which was spanned in space and time, than the older Forster and the younger Chamisso together. They were all members of the Berlin Academy of Sciences in Prussia and highly successful representatives of a

[38] Humboldt, Alexander von: Mes confessions. In: *Le Globe* 7 (janvier–février 1868), p. 182. A German translation is first found in Humboldt, Alexander von: Meine Bekenntnisse (Autobiographische Skizze 1769–1805). In Humboldt, Alexander von: *Aus meinem Leben. Autobiographical Confessions.* Compiled and explained by Kurt-R. Biermann. Munich: Beck 1987, pp. 49–62. My translation draws on the French-language original (O.E.).

translingual Writing-between-Worlds (ZwischenWeltenSchreibens).[39] In them, the highly mobile side of the dynamic configurations of Mobile Prussia shines through.

Humboldt learned from Forster. Perhaps even more than Forster, the author of the *Reise in die Äquinoktial-Gegenden des Neuen Kontinents* was interested in observing phenomena that could only be recorded micrologically, but he never forgot to place his empirically based detailed investigations in a macrological context and thus to relate an observed microcosm to a theoretically thought-out macrocosm. Against the backdrop of his *nomadic* concept of science and knowledge,[40] Alexander von Humboldt's entire life project would certainly be neither conceivable nor understandable without the concept of freedom; in his longing for a comprehensive experience of the world, science and life project went hand in hand.

If Forster's preface to the report on James Cook's *Entdeckungs*reise named the methodological and epistemological premises of an innovative voyage empirically based on field research in the natural and cultural sciences, Alexander von Humboldt, at the end of the eighteenth century, took this conception even further by conceiving and realizing his *Relation historique* as *Forschungs*reise (research expedition) in the modern sense—and in the sense of modernity.

Schooled by Forster's understanding of science and how to do research, Humboldt went far beyond his teacher's views to bring forth a new type of voyage of discovery and a new type of knowledge: Humboldtian transdisciplinary scholarship[41] was in the making and would celebrate its genesis and actual genesis in the *American Travel Diaries*.

There are many good reasons to suspect that the *American Travel Diaries* represent not only the birthplace but the actual monument of this new, epistemologically highly innovative Humboldtian science. And this science was and remained characterized by a nomadic conception of knowledge, not only at the level of thinking and writing, but also of scientific style. Alexander von Humboldt conceived his life neither as a utopia nor as a heterotopia, neither as a distant ideal of his own imagination of nature and culture nor as a longing for the other-place, for a counter-place to his previous life; rather, he constructed his science,[42] but also his entire life as a *vectopia*:

---

[39] Cf. on these concepts Ette, Ottmar: *ZwischenWeltenSchreiben. Literaturen ohne festen Wohnsitz (ÜberLebenswissen II)*. Berlin: Kulturverlag Kadmos 2005. On questions of textualization and literarization, cf. also Görbert, Johannes: *Die Vertextung der Welt. Forschungsreisen als Literatur bei Georg Forster, Alexander von Humboldt und Adelbert von Chamisso*. Berlin: de Gruyter 2014.

[40] Cf. Ette, Ottmar: Nomadisches Denken: Alexander von Humboldts lebendige Wissenschaft. In: Stoyan, Dietrich (ed.): *Bergakademische Geschichten. From the history of the Freiberg Mining Academy told on the occasion of the 250th anniversary of its foundation*. Freiberg: Mitteldeutscher Verlag 2015, pp. 95–102.

[41] Cf. the opening chapter in Ette, Ottmar: *Alexander von Humboldt und die Globalisierung. Das Mobile des Wissens*. Frankfurt a. M./Leipzig: Insel 2009.

[42] Cf. the brief description of the long-term project of the Berlin-Brandenburg Academy of Sciences and Humanities, which began in January 2015, in Ette, Ottmar: Dem Leben auf der Spur. Das Akademievorhaben "Alexander von Humboldt auf Reisen – Wissenschaft aus der Bewegung" ediert Humboldts transdisziplinäre Aufzeichnungen. In: *Die Akademie am Gendarmenmarkt 2015/2016* (Berlin) (2015), pp. 8–13.

as a life out of movement and in constant motion. He embodies not only the mobile Prussian, but at the same time also the mobile Prussia.

There are numerous passages in Alexander von Humboldt's wide-ranging oeuvre in which his nomadic life under the sign of personal as well as scientific freedom presents the often euphorically described movement (understood as *motion* and *emotion*) in condensed form. In this way, his arrival in the world of the American tropics can be understood as an experience of the world that can be interpreted in a scientific, but also in a highly personal and physical way, in the flesh as a literary expression of a vectopia, of a life out of movement and as movement. Thus he wrote in one of his first signs of life from the 'New World', a letter to his brother Wilhelm dated "Cumaná in South America, d. 16 Jul. 1799":[43]

> What trees! Coconut palms, 50 to 60 feet high! Poinciana pulcherrima, with a foot-high bouquet of the most splendid, bright red flowers; Pisange, and a host of trees with immense leaves and fragrant flowers the size of a hand, of which we know nothing. Just think that the country is so unknown that a new genus which Mutis (see Cavanilles iconus, tom. 4) published only 2 years ago, is a 60 feet high far shaded tree. We were so lucky to find this magnificent plant (it had stamens inches long) yesterday. So how great is the number of smaller plants that have yet escaped observation? And what colors of birds, fish, even crabs (sky blue and yellow)! We have been running around like fools until now; during the first three days we cannot determine anything, since one always throws away one object in order to take hold of another. Bonpland assures us that he will go out of his mind if the miracles do not soon cease. But more beautiful than these wonders in detail is the impression made by the whole of this powerful, luxuriant and yet so light, exhilarating, mild plant nature. I feel that I shall be very happy here, and that these impressions will continue to cheer me often in the future.[44]

This place on what is now the Venezuelan coast is not a utopia (nor, of course, a dystopia), but a vectopia: life and science from both external and internal movement. Hardly any other of the many passages in Alexander von Humboldt's extensive oeuvre gives expression with comparable density and intensity to that feeling of happiness that the young European tried to put into words shortly after his arrival in the regions he soon called "South America",[45] soon "America". His experience of the world unfolds here with all its euphoria as a life in a world that has become wider for him, to which the American hemisphere has now been added. Alexander describes to Wilhelm a world of the miraculous: And as an attentive reader of Christopher Columbus, whose surname he carried in the surname of both his and Wilhelm's mother (Colomb), he knew of the long (literary) history of the marvelous in European discourse about the American hemisphere. Humboldt undoubtedly plays with set pieces of this discourse. A scientific as well as personal Eldorado, to be sure: but not in the sense of an other-worldliness, but of a widening and

[43] Humboldt, Alexander von: To Wilhelm von Humboldt. In: Humboldt, Alexander: *Briefe aus Amerika 1799–1804*. ed. by Ulrike Moheit. Berlin: Akademie-Verlag 1993, p. 41.

[44] Ibid., p. 42.

[45] Ibid., p. 41.

expansion. It offered him the chance to continue thinking and to think further[46]—also beyond that new discourse on the New World, as the founder of which, long after the so-called 'Berlin Debate', Alexander von Humboldt may be considered.[47]

The semantics of happiness that permeate the entire letter are omnipresent in their various hues and include the "happiness" that one had in breaking through the blockade of English warships and in the further course of the entire sea voyage—during which he had "worked a lot on the way"[48]—as well as the personal experience of his own happiness and that shared with his French travelling companion Aimé Bonpland. The constellation of movement, work and happiness is omnipresent: the inner movement is closely related to the outer one. Just as the journey seemed to be under a good star, favorable to the whole undertaking, since July 4, 1799, when he "for the first time saw the whole southern cross perfectly clearly",[49] the intense world experience of the new in the 'New World' also produced an intense euphoria in the two European travelers, which in Humboldt's case was always accompanied by a maximum of creative restlessness.[50]

Alexander von Humboldt had arrived at the goal of his great dream, expressed in many letters of his youth: At last, like his friend Georg Forster, he had left the 'Old World' on a journey to regions outside Europe. And yet he was only at the beginning of his American adventure and of that new discourse on the New World of which the *American Travel Diaries* are the birth certificate. Without Humboldt's vectopia, neither they nor Humboldtian science would be conceivable.

Scientific project and life project are intensely interrelated—and here, too, the careful analysis of the *American Travel Diaries* will provide us with numerous new insights. The sense of happiness expressed in the letter to Wilhelm von Humboldt was quite obviously inseparable from the scientific dimension of the expedition. The rapture for which Humboldt searched for ever new words in his letter was the result of a madness, a delocalisation, in which the two Europeans run around like madmen, like "fools",[51] unable to concentrate on a single object of investigation. The madman, in his rapture, refers to the vectopia as the movens of his thinking and writing, of his observing and collecting, of his researching and living. An amazement in the face of "miracles"[52] that do not want to stop has taken hold of the Franco-German research community, comparable to that topical (and tropical)

[46] On the epistemological significance of this widening and expansion, see Ette, Ottmar: Weiter denken. Viellogisches denken/viellogisches Denken und die Wege zu einer Epistemologie der Erweiterung. In: *Romanistische Zeitschrift für Literaturgeschichte* (Heidelberg) XL, 1–4 (2016), pp. 17–60.

[47] On the 'Berlin debate', cf. the second and especially the third chapter in the present book.

[48] Humboldt, Alexander von: An Wilhelm von Humboldt, p. 41.

[49] Ibid., p. 42.

[50] Cf. on restlessness Ette, Ottmar: 'Un esprit d'inquiétude morale'. Vectoricité et économie d'un sentiment intense chez Alexander von Humboldt. In: Espagne, Michel (ed.): *La Sociabilité européenne des frères Humboldt*. Paris: Editions Rue d'Ulm 2016, pp. 47–68.

[51] Ibid.

[52] Ibid.

amazement that seized the European navigators and 'discoverers' during Columbus's first voyages.[53]

But Alexander von Humboldt, in the face of the multitude of all these wonders, did not stop at this *amazement.* The movement, as important as it was as the epistemological basis of his style of thinking, writing and science, did not remain an end in itself, an end in itself, but formed the prerequisite for a not only happy, but happy scientific and life project. Therefore, in his first American letter to his elder brother, which reached him in Spain at the end of October 1799, the not-yet-thirty-year-old by no means persisted in the depiction of a 'foolish', uncontrolled mode of movement, however strong the representation of the feeling of happiness might seem. For the experience of delocalization and dislocation, of external as well as internal movement in the face of the size and colors of those objects that oppose, impose themselves upon, and again elude the Franco-German research team, is immediately coupled with the attempt to relocalize this delocalization with the help of recourse to a science also practiced in America itself, and thus to a *scientific community* in the New World. And thus, above all, to steer it back into controllable and controlled channels.

The reference to the research of the famous Spanish botanist José Celestino Mutis, who had been working in New Granada since 1760 and whom Humboldt was later to visit in the viceroyal capital Bogotá at great expense and with no small advertising effect, in order to compare his "hay"[54] (i.e. his previous plant collections) with this figure central to the Enlightenment in New Granada, blends in that technique of Humboldt's research journey that was of fundamental importance for Humboldt's scientific practice: To undertake a journey not only to the thing to be explored, but also to the explorers there and their findings, not only to the riches of nature, but also to those of the archives and libraries of the Spanish colonial empire. Humboldt's project was purposeful and—also with a view to involving the political power as well as the Creole elites—well thought out: It was only thanks to his clever, diplomatically always skillful choreography that the young Prussian was able to set everything in motion in his favor.

On the scientific level, too, the vectopia of dislocation is still made scientifically productive in the quoted passage itself, in order to redirect it into a choreography of systematic exploration and investigation. Thus the movement by no means disappears, but it is re-channeled: The happy fools are to become happy scientists, who admittedly know how to move in a much broader, wider field. For Humboldt, euphoria is nothing remote from science, but an incentive and a means.[55] Humboldt's

[53] On the relevance of wonder in the history of European expansion, see Greenblatt, Stephen: *Marvelous Possessions. The invention of the stranger: travellers and explorers.* Translated from the English by Robin Cackett. Berlin: Wagenbach 1994, p. 27 f.

[54] Humboldt, Alexander von: *Journey on the Río Magdalena, through the Andes and Mexico.* Part I: Texts. Compiled from his travel diaries and explained by Margot Faak. With an introductory study by Kurt-R. Biermann. Berlin: Akademie-Verlag 1986, p. 93. I will return to the Spanish botanist Antonio José Cavanilles, who is also mentioned in this passage.

[55] Cf. hierzu Kap. 18 ‚Euphorie der Wissenschaft'. In: Ette, Ottmar: *Weltbewußtsein*, pp. 171–183.

happy experience of the world is tied back to a constantly expanding, broadening life. America expands space and time at the same time: the vectopia can be understood as a longed-for movement of experiencing the world that never comes to a standstill.

This always also means that the—in Georg Forster's sense—'micrological' dimension of the individual objects that enrapture the two travelers is increasingly built into macrological contexts. In Humboldt's science, the individual phenomena are always to be related to the 'impression' made by 'the whole of this powerful, luxuriant and yet at the same time so light, exhilarating, mild plant nature'.[56] Humboldt is concerned with the 'total impression',[57] with the perception of the interaction of all forces and factors of nature.

What at first sight appears to be 'out of the ordinary' is thus relationally integrated into a whole, into an overall order within which the individual phenomena, in their significance and functionality, must at the same time be thought apart and put back together, perceived and understood. Humboldtian networking science aims at the overall impression and can be understood as a relational scientific practice that crosses the most diverse bodies of knowledge and disciplines and interconnects them by means of the constant crossings. The 'secret' of its success lies in the vectoriality of these relational crossings, which, thanks to their dynamism, are able to connect everything with everything else.

In this letter from two European researchers in America to a European scholar in Europe, the state of happiness and euphoria of the European scientific subject are undoubtedly linked to the sensual experience of another, non-European world: an experience of the American tropical world that, according to Humboldt's own statement,[58] caused the French travel companion to virtually "take leave of his senses". The vastness of the American continent opens up enormous fields of activity to researchers, which, due to the relative unfamiliarity of these regions, offer enormous potential for innovation for European science. It is this experience of vastness and expansion that regulates the attraction of research objects for research subjects and translates it into a transatlantic asymmetrical economy of the expansion of knowledge.

For the two Europeans conducting field research in America for the first time, the explorable world has fundamentally expanded: it is not so much the 'other', which is directly linked to the 'own' through the same scientific language, as the

---

[56] Humboldt, Alexander von: *Briefe aus Amerika*, p. 42.

[57] Cf. on this, among others, Hard, Gerhard: 'Der Totalcharakter der Landschaft'. Re-interpretation of some passages in Alexander von Humboldt. In: *Alexander von Humboldt: Eigene und neue Wertungen der Reisen, Arbeit und Gedankenwelt.* Wiesbaden: Beilhe 1970, pp. 49–73; Trabant, Jürgen: Der Totaleindruck. Style of Texts and Character of Languages. In: Gumbrecht, Hans Ulrich/Pfeiffer, K. Ludwig (eds.): *Stil. Geschichte und Funktionen eines kulturwissenschaftlichen Diskurselements.* Frankfurt a. M.: Suhrkamp 1986, pp. 169–188; and Kraft, Tobias: From Total Impression to Fractal Representation: the Humboldtian 'Naturbild'. In: Kutzinski, Vera/Ette, Ottmar/Walls, Laura Dassow (eds.): *Alexander von Humboldt and the Americas.* Berlin: Walter Frey—edition tranvía 2012, pp. 144–160.

[58] Humboldt, Alexander von: *Briefe aus Amerika*, p. 42.

'vastness', which transforms the 'world' into a worldwide space of experience. It is only through this '*further*' that a world opens up that does not split into old familiar oppositions between the 'old' and the 'new' world, which have been sufficiently repeated in the *Recherches philosophiques sur les Américains* or the *Histoire des deux Indes*. Even if many remnants of this discourse, which seemed to be able to divide the world into 'own' and 'other' in Cornelius de Pauw, Guillaume-Thomas Raynal or William Robertson, can also be found in Humboldt's discourse, the Prussian cultural and natural scientist clearly goes beyond such constructions of opposition. Cornelius de Pauw and Antoine-Joseph Pernetty may have held opposing views on the significance of the American world and its inhabitants, but they agreed that the Old World hemisphere could be contrasted with the New World hemisphere, irrespective of whether one assessed them positively or negatively, as superior or inferior.

Let us note, then, that since the beginning of European expansion in the first phase of accelerated globalization, this alterity and, even more so, a radical alterization (*othering*) had shaped European discourses.[59] Alexander von Humboldt suggests ways of understanding that no longer insist on and are based exclusively on constructions of the respective other, foreign, non-European, but can be related to an epistemology of networking expansion.

In this respect, the New World—as a scientifically far less explored part of the planet compared to the Old World—offers itself to Humboldt as 'new' in the sense of an expansion, since it holds in store a still incalculable number of previously unexplored plants in the area of the plant world alone. The accent is not on the other, but on the further, which is always to be understood *relationally* and can therefore also be easily related to the previous state of knowledge. This is the source of Humboldt's euphoria, his feeling of happiness in the face of an entire world that still had to be 'discovered' by the sciences and, furthermore, connected with the European one.

His happy world-consciousness[60] is an awareness of a constant widening of the world: far away, but by no means radically separated from that 'meagre sandy nature'[61] into which he had once[62] felt so 'hemmed in' at Schloss Tegel and in Berlin. It is the fullness of a knowledge of the world brought within reach by 'dislocation' that triggers the state of euphoria depicted and deliberately staged by Humboldt at the site of movement of a vectopia: the desire for a wide life, for living (and working) in the widest possible sense.

[59] Cf. Todorov, Tzvetan: *Die Eroberung Amerikas. Das Problem des Anderen. Aus dem Französischen von Wilfried Böhringer*. Frankfurt a. M.: Suhrkamp 1985.

[60] Cf. on this concept of Humboldtian science Ette, Ottmar: *Weltbewußtsein. Alexander von Humboldt und das unvollendete Projekt einer anderen Moderne*. Weilerswist: Velbrück Wissenschaft 2002.

[61] Humboldt, Alexander von: I about myself (my path to becoming a natural scientist and explorer 1769–1790). In: Humboldt, Alexander von: *Aus meinem Leben. Autobiographische Bekenntnisse. Zusammengestellt und erläutert von Kurt-R. Biermann*. Munich: C.H. Beck 1987, p. 38.

[62] Ibid.

Vectopi and Survival

But how, when the concept of *continue to exist (Weiter-Leben)* is supplemented by that of *living on (Weiterleben)* and is primarily related to the temporality and finiteness of one's own life, and thus the risk of one's own death must also be taken into account, which is repeatedly dramatically demonstrated by Alexander von Humboldt's not infrequently daring actions, not only in the published travelogue, but also in the *American Travel Diaries*? How, then, when the traveller in the travelogue is confronted with the possibility of his own death due to the dangers he is threatened with or accepts, and reflects on his continued life?

Such reflections are by no means a rarity or even the exception in travelogues, but are found precisely in the journeys of Europeans to the tropics, which from a European perspective are always both a wealth and a trap, i.e. they shine with their wealth of climatic, geographical or cultural phenomena just as much as with the wealth of tropical diseases, a dangerous animal kingdom or threatening social phenomena (to list here just a few of the stereotypes that repeatedly appear in travelogues to this day). In Alexander von Humboldt's American Travel Diaries, the possibility of his own death is interspersed in a wide variety of ways and reflected upon in written form. Alexander von Humboldt dealt there as much with *continue to exist (Weiter-Leben)* as with *living on (Weiterleben).*

For time and again, Aimé Bonpland and Alexander von Humboldt, together with Carlos Montúfar and their respective local companions and guides, are exposed to deadly dangers at every turn, for example in the volcanic world of the Andes in present-day Ecuador. Not only on Chimborazo, but also on other ascents of the great *Nevados*, death in its omnipresence appears in Alexander von Humboldt's diary again and again in prominent places (and by no means in abstract, theoretical form). For example, the second ascent of Pichincha and especially the risky climb to the crater of Rucupichincha on May 26, 1802 anticipates all the dangers of an ascent in no way comparable to today's safety precautions, but also the traveller's ideas of fear, which we will encounter a few weeks later in the travel diary of the young Prussian on Chimborazo:

> The smell of sulfur announced that we were at the crater, but we doubted that we were above it. A patch of snow barely three feet wide joined two boulders. We walked over this snow in the direction a b. It carried us perfectly. We took two or three steps, the Indian in front and in his Indian pflegma. I was a little on his left behind him, when I saw with a shudder that we were walking on a bridge of snow over the crater itself. [...] And I noticed a blue light between the snow and this stone d. During the second expedition we all saw this blue light in the same hole, it seems to be burning sulphur, because there was no sun to attribute it to a solar reflex. We would have fallen, then, to a depth of 200 toises, and that into the most inflamed part of the crater, and without any one in Quito knowing, if not by our tracks in the snow, what had become of us. I felt myself trembling with terror, and I remember that all I did was to cry out, "Don't move, there is light below," throwing myself on my belly against the rock c, and pulling the Indian back by his ruana (poncho). We believed ourselves safe on this rock c. We discovered that the edge of this rock jutted into the air on all sides except behind us. We had scarcely two toises square to move. We began to examine the danger from which we had saved ourselves. We threw a stone on the snow that lay first to

> the hole through which we had seen the sulphur flame. This stone enlarged the hole, and we ascertained that we had passed over a crevasse (crevasse) between the two rocks b and c, and that a blanket of frozen snow, but of eight inches thickness, had held us.[63]

The infernal fire with its acrid sulphuric stench symbolizes the danger the travelers are in on a precarious snow slab, which Humboldt visualizes with the aid of an attached sketch, directly above the crater opening. He notes a few lines later that he still feels "trepidation as he writes these lines. I see myself hanging again over this horrible maw."[64] A nightmarish scene: the danger to life and limb could hardly be described more impressively. The remembering I and the remembered I step apart and mirror each other in a doubled horror that still lingers and produces afterimages: in the phantasm of one's own death, in the phantasm of one's own disappearance without a trace.

In this interrelation, the traveling I and the writing I intensify in a literarily impressive way not only the danger, lurking as it were among the traveling companions, of being torn into the depths, into a hellish abyss, by a crumbling away of the snow, but also bring this hovering between life and death into a direct connection with their own writing. For Humboldt subsequently not only flashes in a reference to La Condamine's travelogue of a world that could never serve as an "abode for

---

[63] Humboldt, Alexander von: *Reise auf dem Río Magdalena, durch die Anden und Mexiko. Teil I: Texte.* Edited by Margot Faak. *Mit einer einleitenden Studie von Kurt R. Biermann. Zweite, durchgesehene und verbesserte Auflage.* Second, revised and improved edition. Berlin: Akademie-Verlag 2003, p. 203: "The smell of sulfur warned us that we were near the mouth, but we suspected that we were on it. A patch of snow barely three feet wide joined two pieces of rock. We walked on this snow in the direction a b. It carried us perfectly. We took two or three steps, the Indian in front and in his Indian phlegm. I was a little to his left behind him when I saw with a cruel shudder that we were walking on a bridge of snow over the very mouth. I saw that d was a stone supported in the air by the rocks b and c, and I saw a blue glow between the snow and this stone d. We all observed in the second trip this blue light in the same hole, it seems burning sulfur. Because there was no sun to be able to attribute it to a solar reflection. We would have fallen 200 t[oises] deep and in the part of the crater which is the most inflamed and without that in Quito, if not by the traces in the snow, one would have known what we had become. I felt myself shaking with fear and I remember that all I did was to cry out: "Quiet, light from below", throwing myself on my stomach against the rock and pulling the Indian by his roan (Poncho). We thought we were safe on this rock c. We discovered that on all sides except behind us, the edge of this rock was in the air. We had hardly two square toises to move. We began to examine the danger from which we had escaped. We threw a stone on the snow closer to the hole through which we had seen the sulfur flame. This stone enlarged the hole and we were reassured that we had stepped on a crevasse between the two rocks b and c and that a layer of frozen snow, but hardly 8 inches thick, had supported us. We believe that this crevasse goes only up to e f, because from there to the left we could not sink the snow and we imagine that the rock c, holds there to the rock b. We passed there in this trip and in the following one without danger and it is the safest way to reach the rock which forms a gallery above the crater." Reference is made here to the original of Humboldt's *Amerikanischen Reisetagebüchern* VII bb/c, 11v–12r as well as the easily consultable digital copies of the original; digital copy 11v: http://resolver.staatsbibliothek-berlin.de/SBB000152B400000499; and digital copy 12r: http://resolver.staatsbibliothek-berlin.de/SBB000152B400000500.

[64] Ibid.: "[…] mais en même temps plus désagréable. Je me sens étouffé (ansio[so]) en écrivant ces lignes. Je me crois encore suspendu sur ce gouffre affreux." Cf. also digital copy of Diary VII bb/c, 12r.

living creatures,"[65] but also immediately afterward and not coincidentally refers to Milton's *Paradise Lost*, which he interestingly attributes to Pope in the excitement of his writing. Above all, however, he establishes the relationship with his own writing, which not only presents and represents the horror of impending death, but at the same time seeks to banish it and open it up to the imagination of man and to literature. For it is from literature that Humboldt draws the reservoir of images he uses intertextually to understand and stage his own situation and that of his survival. At the same time, literature points to a dimension that goes far beyond his own life: it *is* the medium of continued life. In it, the work of Dante, Milton or Pope is still alive centuries later.

Without a doubt, the writing down of the fear of death is first and foremost the sign of one's own survival: a sign of life to others, to readers, to oneself, of having survived, of still being alive. For anyone who can describe and write down a danger must first of all have survived it. Thus writing becomes a sign of (one's own) survival, indeed it is able to transform itself into a *knowledge of survival*.[66] In this very elementary sense, the travel diary itself is a sign and proof not only of a will to survive, but also of a knowledge of survival that connects the traveller and the writer, the one who is remembered and the one who remembers, the one who exposes himself to danger and the one who translates this mortal danger into the life knowledge of literature, into an aesthetic unity that is always precarious and endangered. Yet the question expressed in the above quotation, what would have happened *after the death* of the travel group if, in the search for the missing, one had at best found traces still lost in the snow, remains open for the time being.

The Humboldtian phantasm against which the writing repeatedly struggles is Humboldt's own disappearance from the surface of the earth without leaving any traces. This phantasm is found variously and consistently throbbing in his *Amerikanischen Reisetagebüchern*. Here, one's own writing becomes an antidote, a food and a means of survival, insofar as it aims to leave traces, not simply to disappear. The daily writing down of one's experiences, which has become a routine and a ritual, is able to *record* one's own life in writing and to translate it into a potential reading of others, of contemporaries as well as of posterity. Life is thus translated into a reading that ensures the survival of the traveling as well as the writing self, and transforms writing into that always precarious bridge that straddles the cleft, the fall into nothingness, in lines pressed tightly together. But is this only about survival?

Humboldt's writing undoubtedly moves between the hell of the crater, of his own disappearance, and the paradise of literature, which proclaims *and* depicts, presents *and* represents, as it were, an 'eternal' life. Hardly anything occupied and depressed Humboldt's imagination more enduringly than the images, variously fleshed out and rising in his diary, of (his own) tracks leading to nothingness: be it a crash into the crater, a capsizing on the Orinoco, or a shipwreck of the kind he later feared on

[65] Ibid.: "On croit voir un monde [two words deleted] détruit et sans espérance de pouvoir jamais servir de demeure à des êtres organisés."

[66] Cf. on this concept Ette, Ottmar: *ÜberLebensWissen I-III*. Drei Bände im Schuber. Berlin: Kulturverlag Kadmos 2004–2010.

crossing to the USA. It is the phantasm of his own disappearance without a trace that he creates literarily while writing, and to which he defiantly opposes this writing in order precisely not to disappear without a trace. Through the continuation of the diary, the continuation of the journey, of the writing, of his own life is literally *recorded*—even if in the diary only from day to day.

In all the adaptations that Alexander von Humboldt gave to his attempt to climb the summit of Chimborazo, the life of the travellers moves almost obsessively on a knife's edge: on that *cuchilla*, which—as Humboldt knew very well—in Spanish stands not only for the narrow ridge of rock, but in the literal sense for the dissecting knife, the knife blade and the butcher's axe. So we read in the lines of the diary always rising more or less in writing direction:

> We came upon a narrow ridge, a very peculiar cuchilla. The path was barely 5–6 inches wide, sometimes not 2 inches. The slope to the left was of frightening steepness and covered with snow frozen to the surface. To the left there was no atom of snow, but the slope was covered with large boulders. One had the choice of preferring to break one's bones striking these rocks, from which one would have been beautifully received at a depth of 160–200 toises, or to roll over the snow to the left into a much deeper precipice. The latter fall seemed to us the more horrible. The frozen crust was thin, and one would have been buried in the snow with no hope of ever resurfacing. For this reason we always leaned our bodies to the right.[67]

The phantasm of communal death is sketched here alternatively in two directions. Quite obviously we encounter again the image of disappearance, of dissolving into or under matter, following Humboldt. The preferred inclination to the right stands for the mode of death preferred by the travelers, which would at least have ensured that the fallen person would not have disappeared without trace from the surface of the earth and sunk into the deep soft snow without ever resurfacing. As in the hellfire of the crater, he would have been hidden from any trace under the icy blanket of snow.

The fall to the right, on the other hand, would ensure that the rocks on which the traveller would have to hit at the end of his fall would at least still "beautifully receive" the human body and thus preserve it for posterity—albeit shattered. Otherwise—and this was Humboldt's idea—the explorer's body as well as his diary would have disappeared forever under the blanket of snow, untraceable. A fall in either direction had to be fatal: *Survival* was out of the question. But in falling onto

[67] To give a textual impression of the American Travel Diaries at this point, we will not refer to the edition edited by Margot Faak, but to a precise transcription of the digital copies: "Nous trouvames une file une cuchilla très curieuse. Le chemin avait a peine 5–6 pouces de large quelquefois pas 2 po[uces]. A gauche la pente etait d'une rapidité effrayante et couverte de neige gelée (croutée) à la surface. A droite il n'y avait pas un atome de neige mais la pente etait semée (couverte) de grandes masses de roches. On avait a choisir s'il [letter deleted] valait mieux se briser les membres en tombant contres ces rochers où l'on aurait eté bien recu à 160–200 t.[oises] de profondeur, ou si a gauche on voulait rouler sur la neige a un [word deleted] abime beaucoup plus profond. La derniere chute nous parut la plus affreuse. La croute gelée était mince et on se serait enterré dans la neige sans esperance de revenir au jour. C'est pour cela que nous laissions toujours notre Corps penché a droite."

the rocks, the travel diary would be saved from disappearance: it is not a matter of survival here, but of a *continuation of life* after death in writing.

Thus, the movement opens on a knife's edge towards two possible ways of death, only one of which is irrevocable, but the other—thanks to the presence of the writing preserved by the diary—is ultimately indelible and allows for a continuation of life. Mere survival ensures only the continuation of life up to that point at which physical life inevitably finds its terminus. But continuing to live goes beyond this apparent end point of one's life and makes one's life permanent until further notice. Continuing life aims at that which clearly goes beyond space and time, beyond one's own physical limitations. Living on does not mean eternity: it is about a life beyond death, admittedly a living on until further notice: as long as there are readers who are willing and able to decode the Scriptures. Understood in this sense, the written continuation of life is time deferred.

Edition projects are—this should be noted here in passing—projects that aim to keep a writing, a writing and a thinking alive far beyond the physical death of the author (as well as the editor) and to enable it to live on. Furthermore, an editing project is not about 'dead' letters, but about the chance to secure and keep *alive* the continued life of a thought (and a thinker), so that in polylogue with the living it can always take on *further* meanings beyond what was thought while alive. For what is written, the text before us, is not reducible to what was thought at a particular time. It literally lives on.

Such a living on does not aim at preservation for the sake of preservation, but leads to a further living of this thinking and thus to a further, constantly expanding life of the living. In this sense, it is not a matter of 'naked' survival,[68] but of a *knowledge of living on (Weiterlebenswissen)* that is always vectorially inscribed in the act of writing as hope, always as a *prospective* trace. Alexander von Humboldt's reflection on the different modes of death proves how much his own writing longed for traces that could lead into the future and secure a continued life for his experience of the world.

Adelbert von Chamisso (1781–1838)

Vectopia of World Experience

In Adelbert von Chamisso's letter of February 18, 1810 from Paris to his friend, the publicist Julius Eduard Hitzig, who five years later was to play a decisive role in the success of Chamisso's application to take part in the Russian circumnavigation of the globe, the great Prussian poet, who had fled France during the French Revolution

[68] Cf. on this Agamben, Giorgio: *Homo sacer. Die souveräne Macht und das nackte Leben.* Aus dem Italienischen von Hubert Thüring. Frankfurt a. M.: Suhrkamp 2002.

at the age of eleven,[69] reported with great admiration on Alexander von Humboldt's numerous activities: "Such activity, speed and firmness have never been seen before".[70] He was busy with the publication of his travel work, was moreover often at court and at the same time was preparing his "new forthcoming excursion".[71] Humboldt wanted to sail to the Cape of Good Hope for observations and from there travel on to India and Bengal, in order to then explore Tibet and the interior of Asia.

These lines of Chamisso's, in a condensed form and with great admiration, throw a telling light on that urge to action and movement of the author of the *Ansichten der Natur*, on that vectopia which led Humboldt to incessant changes of place. Not only on a global scale, by the way, but also in the French capital itself: Humboldt "spent his nights at the observatory" and inhabited no less than "three different houses."[72] The rhythm of Humboldt's life is not only likely to impress his contemporaries.

Even though Alexander von Humboldt was never able to carry out his long-awaited journey to India and Tibet due to the hostile attitude of the colonial power Great Britain, which not without reason feared the colonial criticism of the Prussian, and was only able to carry out his journey to Asia in 1829 with the kind and calculated help of the Tsar in his empire, this *movement picture* of the scholar has its full justification. Humboldt rarely kept it in one place.

For his part, Adelbert von Chamisso saw in Humboldt a lifestyle that he was to emulate in later years. It was not without reason that Humboldt asked the "circumnavigator"[73]—as he put it in a letter to Chamisso at the beginning of 1828—for detailed information in his letters and reported in a letter probably dated May 16, 1836 of his delight at how much "your life story, your voyage, your so eloquently noble and firm picture had made such a deep, benevolent impression on the honourable Crown Prince".[74] Chamisso's travelogue, which had just been published, had been read aloud—probably by Humboldt himself, who also served at court as a chamberlain in his function as a reader—and had radiated the charm of a new world

[69] On the biographical background of the poet and naturalist cf. among others the biographies published with a certain regularity and documenting the further growing interest in chamisso over four decades by Freudel, Werner: *Adelbert von Chamisso. Leben und Werk.* Leipzig: Philipp Reclam jun. 1980; Fischer, Robert: *Adelbert von Chamisso. Weltbürger, Naturforscher und Dichter.* Vorwort von Rafik Schami. Mit zahlreichen Abbildungen. Berlin/München: Erika Klopp 1990; Arz, Maike: *Literatur und Lebenskraft. Vitalistische Naturforschung und bürgerliche Literatur um 1800.* Stuttgart: M&P Verlag für Wissenschaft und Forschung 1996; Langner, Beatrix: *Der wilde Europäer. Adelbert von Chamisso.* Berlin: Matthes & Seitz 2009.

[70] Chamisso, Adelbert von: Letter of 18.02.1810 to Hitzig. In: *Adelbert von Chamisso's Werke.* 5th, enlarged edition. Vol. 5: *Life and letters of Adelbert von Chamisso.* Edited by Julius Eduard Hitzig. Berlin: Weidmann'sche Buchhandlung 1842, p. 276.

[71] Ibid.

[72] Ibid.

[73] Humboldt, Alexander von: Letter to Adelbert von Chamisso [probably early 1828]. Staatsbibliothek zu Berlin, Nachlass Adelbert von Chamisso, acc. ms. 1937, 183.

[74] Humboldt, Alexander von: Brief an Adelbert von Chamisso [vom 16.05.1836], Staatsbibliothek zu Berlin, Nachlass Adelbert von Chamisso, acc. ms. 1937, 183.

drama[75] through the "individuality of the presentation", as the poet Chamisso understood in an extraordinary way how to be able to "write prose in an unbiased, simple and free manner".[76]

Humboldt had an extraordinary appreciation for the specifically literary aspects of that experience of the world that was so artfully expressed in the account of Chamisso's great circumnavigation of the globe to be discussed in the following. It was not for nothing that Chamisso had dedicated a remarkable poem to the account of a Guahiba Indian woman hunted down and tortured by Christians, which he had found in Humboldt's American travelogue of the *Relation historique.*[77] For his part, Humboldt probably found in Chamisso's travel-literary writing the idea that he had described in his 'Preface to the Second and Third Editions' of his *Ansichten der Natur*, dated March 1849, as the "combination of a literary and a purely scientific purpose".[78]

In his travelogue, Adelbert von Chamisso had not failed to express his sincere admiration for Alexander von Humboldt at various points. Thus, in his equally elaborate and complex account of the second Russian circumnavigation of the world, which he undertook from 1815 to 1818 on board the Russian brig 'Rurik' under the command of Captain Otto von Kotzebue, he wrote with great respect for the Prussian who had long since achieved world fame:

> Don Jose de Medinilla y Pineda had known Alexander von Humboldt in Peru, from whence he came to these islands, and was proud to have lent him his own hat once, when the latter sought one to appear at the court of the Viceroy. Later, in Manila, the capital of the Philippines, which has always been in lively contact with the New World, we have often heard the world-famous name of our compatriot mentioned with veneration, and we have met several gentlemen, especially ecclesiastics, who boasted of having seen or known him.[79]

However, not only the world fame, but also the knowledge and especially the ability to think together, which was so characteristic for Humboldt's thinking, writing and scientific style, was emphasized again and again in Chamisso's *A Voyage Round the World with the Romanzoffian Discovery Expedition in the years 1815–1818 on the brig Rurik*, succeeded a Humboldt,

> to survey and illuminate the fragments of local meteorological observations, which exist only as meagre contributions to a physical geography, and to bring them under a law, to

---

[75] Ibid.

[76] Ibid.

[77] Cf. on this poem Lamping, Dieter: 'Ein armer unbedachter Gast'. Adelbert von Chamisso's Intercultural Poetry. In: Chiellino, Carmine/Shchyhlevska, Natalia (eds.): *Bewegte Sprache. From 'Gastarbeiterdeutsch' to intercultural writing.* Dresden: Thelen 2014, esp. pp. 20–25.

[78] Humboldt, Alexander von: *Ansichten der Natur mit wissenschaftlichen Erläuterungen.* Tübingen: Cotta 1808, p. 9.

[79] Chamisso, Adelbert von: Reise um die Welt mit der Romanzoffischen Entdeckungs Expedition in den Jahren 1815–18 auf der Brigg Rurik Kapitän Otto von ‚Kotzebues'. In: Chamisso, Adelbert von: *Sämtliche Werke.* Bd. II. München: Winkler 1975, p. 224.

> draw isothermal lines across the globe, to subject a hypothesis for explaining the phenomena to the test of naturalists.[80]

The traces of Alexander von Humboldt's veneration, but also his long shadow, are easily noticeable in explicit or implicit form in many places of Chamisso's *A Voyage Round the World*, In his own investigations, for example, he also drew on the findings of the plant geography devised by Humboldt[81]—which the author of the *Vues des Cordillères et Monumens des Peuples indigènes de l'Amérique*[82] cited by Chamisso commented on in a letter to Chamisso, not without the mischievousness so typical of him, that Chamisso had after all "withdrawn many a plant geographic from him" in his "general travel observations".[83] The references to the manifold connections between Humboldt and Chamisso could easily be multiplied.

But Adelbert von Chamisso's *A Voyage Round the World* had a completely different character than Alexander von Humboldt's multivolume (and, incidentally, prohibitively expensive) American travel work. This is already due to the completely different nature of the voyage, since Chamisso—like Georg Forster and his father Reinhold—was not his own master, unlike Humboldt, but had to obey orders on a Russian warship. The resulting differences can already be seen in the first pages of the *A Voyage Round the World*.

The difficulties that the rather headstrong Chamisso had already encountered during the voyage with the younger but very experienced captain of the second Russian circumnavigation of the world (after Admiral Krusenstern's voyage) continued after his return to Europe. He was forbidden by the highest authorities to publish his voyage report in advance under his own name, so that the scientific results he achieved did not appear until 1821 in the third volume of Otto von Kotzebue's Reisebericht.[84] As early as 1819 Chamisso had been working intently on his report; but it was not until 1836 that the travel work could appear under his name. It was clearly divided into the two parts 'Diary' and 'Remarks and Views'. The rivalry with the official travel report published by the Russian Admiralty under the name of Otto von Kotzebue, the son of August von Kotzebue, was palpable in Chamisso's formulations.

Even though the paratextual title elements of both 'Reise um die Welt' and 'Ansichten' may be obvious allusions to Georg Forster's *Reise um die Welt*, but also to *Ansichten vom Niederrhein* and Humboldt's *Ansichten der Natur*, neither the

---

[80] Ibid., vol. II, p. 472 f.

[81] Cf. inter alia ibid., vol. II, p. 308.

[82] Cf. the German edition of Humboldt, Alexander von: *Views of the Cordilleras and Monuments of the Native Peoples of America*. Translated from the French by Claudia Kalscheuer. Edited and with an afterword by Oliver Lubrich and Ottmar Ette. Frankfurt a. M.: Eichborn 2004.

[83] Humboldt, Alexander von: Letter to Adelbert von Chamisso [probably dated 16.05.1836]. Staatsbibliothek zu Berlin, Nachlaß Adelbert von Chamisso, acc. ms. 1937, 183.

[84] Cf. Federhofer, Marie-Theres: Lokales Wissen in den Reisebeschreibungen von Otto von Kotzebue und Adelbert von Chamisso. In: Kasten, Erich: *Reisen an den Rand des Russischen Reiches: Die wissenschaftliche Erschließung der nordpazifischen Küstengebiete im 18. und 19. Jahrhundert.* Fürstenberg/Havel: Kulturstiftung Sibirien 2013, p. 131.

composition and structure nor the tone adopted by Chamisso can be placed in the vicinity of the so successful models Forster and Humboldt. The Franco-Prussian poet went his own way here in many respects. It was not for nothing that Humboldt, in his aforementioned letter of May 16, 1836, spoke of Chamisso's—and the poet himself had chosen[85] this term—"life story".[86]

If in the second part of the work the 'Remarks and Views' offered in essence the scholarly results already printed (and now partly updated) in the third volume of Kotzebue's account, the 'Diary', the first part of Chamisso's *A Voyage Round the World*, further developed the genre-specific proximity between travel-literary and autobiographical forms of writing, so that the journey can also be read as a journey through life in an autobiographical mode of reading.

At the same time, the traveller, who had begun his career as a natural scientist only a few years before the start of his great journey, in 1812, with the decision to study medicine and natural sciences at the newly founded Berlin University, readily admitted his deficits and weaknesses, quite unlike a Humboldt. Where Humboldt only admitted in a private letter to Wilhelm the still lacking knowledge of many plants in order to immediately supply scientific terminology even there, Chamisso disclosed the respective state of his own knowledge in his published work.

Thus he had felt a stranger on board from the very beginning and had clearly recognized his own limitations, as the voyager, who was on the high seas for the first time and who had always suffered from the swell, had already indicated at Plymouth with a view of the English coast:

> The tide rises to twenty-two feet on the transitional limestone and clay-shale cliffs; and at low tide the richest, most wonderfully mysterious world is revealed to the naturalist's eye. I have not since encountered anywhere a beach equally rich in tangs and sea-worms. I recognized almost none of these animals; I could not find them in my books, and I was indignant at my ignorance. It was not until later that I learned that most of them must really be unknown and undescribed. In the course of the journey I missed many a thing in this way, and I record it here with care for the lesson of my followers. Observe, you friends, collect and store for science what comes into your sphere, and do not let the opinion err: this and that must be known, and only you do not know about it.[87]

Observing, collecting and storing for science is described here as a long learning process within which the researcher need not be ashamed of his temporary ignorance. For Chamisso, this learning process is part of a life that is not only scientific, which on this journey around the world, on this journey through life, is under the sign of an experience that has as its object "the richest, most wonderfully enigmatic world".[88] In the 'Diary' this experience of the world is repeatedly brought into focus

[85] Cf. Chamisso, Adelbert von: *Reise um die Welt*, vol. II, p. 9.

[86] Humboldt, Alexander von: *Letter to Adelbert von Chamisso* [probably dated 16.05.1836]. Staatsbibliothek zu Berlin, Nachlass Adelbert von Chamisso, acc. ms. 1937, 183.

[87] Chamisso, Adelbert von: *Reise um die Welt*, vol. II, p. 24.

[88] Ibid.

in the play between the remembering and the traveling self. It ultimately leads Chamisso to *another* kind of knowledge.

Knowledge on the Threshold of Survival

If Adelbert von Chamisso already began to dream of a world tour as a naturalist when he began his studies in 1812,[89] then the vectopia, habitus and rhythm of life of an Alexander von Humboldt—as we have seen—may not have been entirely uninvolved in this project. And if in his global success *Peter Schlemihl's Wondrous Tale*, written in 1813 at his retreat in Kunersdorf,[90] on the 'Musenhof' of the von Itzenplitz family, he[91] equipped his hero and naturalist with "seven-league boots[s] on his feet", then the mighty strides of his Schlemihl have something to do with that "kind of philosophical rage"[92] of which Johann Gottfried Herder spoke in 1774. But also with those "quick steps, where the whole earth will be revealed to the European spirit of research and close every gap in our empirical sciences",[93] as Georg Forster aptly formulated it for his epoch in 1791.

In his travelogue from 1836, Chamisso signals with tremendous intensity the new experience of the world that comes to him as a naturalist—just as his Peter Schlemihl only slowly begins to become aware of the effect of his seven-league boots:

> I didn't know what had happened to me, the freezing frost forced me to quicken my steps, I only heard the roar of distant waters, one step and I was on the icy shore of an ocean. Innumerable herds of seals plunged into the floods before me, rushing. I followed this shore, I saw bare rocks again, land, forests of birch and fir, I walked straight ahead for a few more minutes. It was stifling hot, I looked around, I was standing among beautifully built rice fields under mulberry trees.[94]

It is certainly no coincidence that at the end of the nineteenth century the Cuban poet and essayist José Martí, in his epoch-making and epoch-embodying essay *Nuestra América*[95] in 1891, in the midst of the third phase of accelerated globalization, mentioned and unpacked Chamisso's seven-league boots in order to refer them, with regard to the expanding USA, to that giant, the United States of America, which would soon put its boots on the chest of southern America. For the speed with

[89] Cf. Federhofer, Marie-Theres: *Lokales Wissen in den Reisebeschreibungen von Otto von Kotzebue und Adelbert von Chamisso*, p. 120.

[90] Cf. in detail Sproll, Monika: *Adelbert von Chamisso in Cunersdorf.* Frankfurt (Oder): Kleist-Museum 2014.

[91] Chamisso, Adelbert von: Peter Schlemihl's wondrous journey. In: Chamisso, Adelbert von: *Sämtliche Werke*, vol. 1, p. 60.

[92] Herder, Johann Gottfried: *Also a Philosophy of History for the Education of Mankind*, p. 89.

[93] Forster, Georg: *Die Nordwestküste von Amerika, und der dortige Pelzhandel*, p. 390.

[94] Chamisso, Adelbert von: *Peter Schlemihl's Wondrous Tale*, p. 60.

[95] Cf. on this Ette, Ottmar: José Martís *Nuestra América* oder Wege zu einem amerikanischen Humanismus. In: Röseberg, Dorothee (ed.): *El arte de crear memoria. Festschrift on the occasion of the 80th birthday of Hans-Otto Dill.* Berlin: trafo Wissenschaftsverlag 2015, pp. 75–98.

which planetary space is traversed, visible in the above passage, stems from that experience of a worldwide acceleration which had seized and changed the earth in the second phase of accelerated globalization in the eighteenth century. The world had changed irreversibly, because the *experience* of this world had changed fundamentally in the sign of an accelerated movement worldwide.

If the series of Russian circumnavigations still followed the model of those voyages of discovery (*Entdeckungs*reisen) of Cook, Lapérouse, or Bougainville, with the help of which the leading powers of the second phase of globalization had laid claim to a global leadership role as well as to vast land areas to be colonized. Despite their delay of several decades, they can still be counted as part of the acceleration that was postponed in this phase and whose protagonists included the French-born immigrant from Prussia who sailed around the world on a Russian warship under German-Baltic leadership, not least for the benefit of the Russian-American Trading Company. Russia had embarked on the long road to world power; and Chamisso, weary and already near death, published his *A Voyage Round the World* as an account that reflected this road to world power, but was at the same time a Schlemihlian journey around the planet as well as around his own life. Especially in his 'Diary' he developed for it a special, poetically condensed language that unfolds its very own magic.

If Georg Forster and Alexander von Humboldt made use of the two great languages of the second phase of accelerated globalization, which were rapidly spreading around the world, by having the former publish his *A Voyage Round the World* first in English and only a little later in German, while the latter presented most of his *Voyage aux Régions équinoxiales du Nouveau Continent* in French and published only certain parts in German, Chamisso, who was born in France and fled to Berlin at the age of eleven with his aristocratic family, which was rebelling against the French Revolution, chose the German language for his travelogue, without, of course, forgetting to draw attention effectively to the multilingual world on board the Russian brig.

Not without a humorous wink, he repeatedly described his own highly original linguistic-cultural position within the microcosm of the 'Rurik', the world on board the ship. Just as mischievously as enigmatically, this linguistic microcosm is then brought into contact with the macrocosm outside the ship: a multilingual world, not reducible to a single language, not to a single logic, emerges under the pen of this well-travelled and mobile Prussian.

For example, at the end of his stay in the Canary Islands, he noted in his travel report:

> At first on Tenerife, as later everywhere in the whole world, the inquisitive people with whom I came into closer contact as an inquisitive person took pains to study the Russian national character in me, the Russian, who was nevertheless only a German, and as a German actually a born Frenchman, a Champenois.[96]

[96] Chamisso, Adelbert von: *Reise um die Welt*, vol. II, p. 42.

The oscillation between belonging and not belonging to a nation runs like a thread through Chamisso's account of a circumnavigation of the world, which takes place at the very beginning of the nineteenth century and thus of the century of great nationalisms.

Adelbert von Chamisso had already pointed out on the first pages of his biographical travelogue that in 1813, as a native Frenchman and former Prussian officer, he was not "allowed to take an active part"[97] in the great national uprising against Napoleon's hegemony over Europe: "I no longer had a fatherland, or not yet a fatherland",[98] all these events "repeatedly tore me apart in many ways".[99] The biographical parallels to Humboldt's or Forster's oscillation between Germany and France are evident in Chamisso's work. Chamisso stressed that he had taken refuge in Prussian Kunersdorf and in writing his *Peter Schlemihl in order* to "disperse himself and to amuse the children of a friend".[100] Weren't the seven-league boots the best possible literary realization of a vectopia that was to be vividly realized a short time later with his *A Voyage Round the World*?

The microcosms of Kunersdorf and the 'Rurik' also served, however, to come to an agreement with himself on a further, expanded understanding of his time beyond a nationalism that was burgeoning everywhere. As a migrant, was he not the living example of a life that, as in the case of Georg Forster or Alexander von Humboldt, could not and was not allowed to be limited to a nation, to a nationality, to a particular national feeling? Wasn't he himself—and in this quite similar to Humboldt—the best example of that *mobile Prussia* that only with the Napoleonic era began to suppress the history of its rise in the nationalist frenzy? And was not the cosmopolitan Prussia that Adelbert von Chamisso stood for a Prussia in which views beyond the national were reflected in a fascinating way?

On the way from Berlin to Copenhagen in Denmark, where he was to join the expedition under Otto von Kotzebue and become part of the internationally equipped research team as an official participant in the Russian circumnavigation of the world, Adelbert von Chamisso found, not without a certain astonishment, in himself "the gift at all" of finding himself "at home everywhere at once".[101] Thus, even before embarking on his first and only world tour, he began to settle down beyond the national, and thus to be less a stranger who feels himself a stranger everywhere than a nomad to whom movement everywhere offers a dwelling-place and a life without a fixed abode as a movable 'home'. He does not miss either of his fatherlands; and of his international circle of friends in Copenhagen he knows to tell us that there he "passed perhaps the most cheerful and merry days of my life."[102] Adelbert von Chamisso stands for a Prussia that cheerfully incorporates its diverse origins and knows how to translate them into no less diverse futures.

[97] Ibid., vol. II, p. 11.

[98] Ibid.

[99] Ibid.

[100] Ibid.

[101] Ibid., vol. II, p. 16.

[102] Ibid., vol. II, p. 17.

With Georg Forster as well as with Alexander von Humboldt, Adelbert von Chamisso shares the experience of translingual writing, consequently of an *écriture* that also moves beyond one's own mother tongue and repeatedly oscillates between different languages. It may have been precisely this life and writing without fixed linguistic and national coordinates that allowed him, when observing, collecting and storing data on the cultures of other peoples, to always include reflection on his own, European point of view and, by means of this critical self-examination, to develop a view of indigenous cultures that was far less marked by prejudice than was the case, for example, in Otto von Kotzebue's official report of the second Russian circumnavigation of the world.[103]

Only a few years earlier, living in the circle of Germaine de Staël, Adelbert von Chamisso had been close to despair, believing that he could not find his home anywhere:[104]

> My homeland. I am French in Germany, and German in France, a Catholic among Protestants, a Protestant among Catholics, a philosopher among religious people, and a fool among people without prejudice; a man of the world among scholars, and a pedant in the world, a Jacobin among aristocrats, and a nobleman among democrats, a man of the Old Regime, etc. I am nowhere to be found, I am everywhere a stranger—I would like to embrace too much, everything escapes me. I am unhappy… Since tonight the place is not yet taken, allow me to go and throw myself head first into the river…[105]

The account of his *Voyage Round the World*, on the other hand, shows us a Chamisso who does indeed look back on his own brokenness, who does not fade it out or repress it, but who has, as it were, transformed it into a dynamic movement. Neither a heterotopia nor an atopia, and certainly not a utopia or dystopia, freed him from his specific foreignness: rather, it was a vectopia that allowed the stranger everywhere to become a nomad (in contrast to Forster and Humboldt, who were certainly only temporary), who knew how to relativize the prevailing forms and norms of life and tried to create his own mobile home, participating in the most diverse languages and logics. It was certainly no coincidence that Chamisso, in his last scientific publications, dealt with the structure and composition of the Hawaiian language until the end of his life. For even for him, the world could not be grasped and understood solely from the perspective of a single language.

The fixed borders have rather become mobile transitions, the mutual exclusion mechanisms rather translation processes, which bring the foreigner closer to a nomadic, multi-logical knowledge. This process of change and transformation is evident not only in the heavily autobiographically tinged passages of Chamisso's *A Voyage Round the World*, but also in the (scholarly) nature of the multi-perspectival

---

[103] Cf. Federhofer, Marie-Theres: *Lokales Wissen in den Reisebeschreibungen von Otto von Kotzebue und Adelbert von Chamisso*, p. 133.

[104] Cf. Fischer, Robert: *Adelbert von Chamisso. Weltbürger, Naturforscher und Dichter.* Foreword by Rafik Schami. With numerous illustrations. Berlin/Munich: Erika Klopp 1990, p. 98.

[105] Chamisso, Adelbert von: *Leben und Briefe.* Edited by Julius Eduard Hitzig. Leipzig: Weidmann'sche Buchhandlung 1942, p. 271.

engagement with the objects he examines. In other words, the literary travelogue presents us with a very profound transformation from an experience of the world that has become increasingly polylogical.

When Chamisso, as a great writer, not only prefaces his own narrative with a *mise en abyme of* narrating a voyage around the world,[106] but also early on introduces the formula of a knowledge of life as knowledge of life in life and for life- "surely it must be the same for him on the brig 'Rurik' as it is in the world in general, where only life teaches life"[107]—at the same time he establishes the relation between that "small world",[108] into which the "nutshell" of[109] his ship has now "pressed"[110] him, and that comprehensive world, which stand to each other as in a relation of microcosm and macrocosm.

With this relationship between micro- and macro-structure, which we had encountered before, the *moving* basis of his constantly changing perspectivism on the world is highlighted, a movement that corresponds to the changing contours and contexts of his life. This allows the reading audience to understand that the view of the 'wide' world is a very specific one, determined by the context of the ship itself, without this moving perspective becoming a 'natural' one that can no longer be questioned.

On a ship—and even more so on a Russian warship—one cannot simply avoid other 'co-inhabitants';[111] the resulting problems are those of conviviality, of a necessary coexistence in a limited available space. In this respect, it is not surprising that Chamisso repeatedly deals with questions and challenges concerning conviviality between different nations, languages or cultures, conviviality between military men and scientists, but also between Europeans and indigenous peoples.

For just as there are constant conflicts in the microcosm of the ship, subject to a highly hierarchized mode of procedure, so too the forms as well as the norms of conviviality are constantly examined by Chamisso in the light of conflicts, indeed of catastrophes, as they occur at the level of power relations between, say, colonial masters and the colonized. Adelbert von Chamisso's *A Voyage Round the World* poses the question of living together on a nutshell that is nothing other than the world. The idea of the 'Spaceship Earth'[112] is far older than manned space travel with its images of the Lonely Planet.

Chamisso consciously seeks, where at all possible, to distance himself from any arrogance of the so-called 'civilized' towards the so-called 'savage'[113] and to adopt

---

106 Cf. Chamisso, Adelbert von: *Reise um die Welt*, vol. II, p. 15.

107 Ibid., p. 20.

108 Ibid.

109 Ibid.

110 Ibid.

111 Ibid., vol. II, p. 37.

112 Cf. Grober, Ulrich: *Die Entdeckung der Nachhaltigkeit. Kulturgeschichte eines Begriffs*. Munich: Antje Kunstmann 2010, p. 227 f.

113 Cf. on this regulative of European world appropriation Bitterli, Urs: *Die 'Wilden' und die 'Zivilisierten'. Grundzüge einer Geistes- und Kulturgeschichte der europäisch-überseeischen*

a *wider* perspective that escapes the standard of alterity and alterization. Thus he resolutely states:

> I take this opportunity here also to lodge a solemn protest against the appellation 'savages' in its application to the South Sea Islanders. I like to associate, as I can, certain concepts with the words I use. A savage, to me, is the man who, having no fixed abode, cultivation of fields, or tamed animals, knows no other possession than his weapons, with which he subsists by hunting. Where depravity of manners may be blamed on the South Sea Islanders, such does not seem to me to testify to savagery, but rather to super-morality. The various inventions, the coin, writing, etc., which are suitable for measuring the various levels of morality at which the peoples of our continent find themselves, cease, under such changed conditions, to provide a yardstick for these insularly segregated families of men who, under this blissful sky without yesterday or tomorrow, live for the moment and for pleasure.[114]

The critical reflections on language merge into critical reflections on civilization, which, despite all romanticizing coloration—the first person plural indicates it—do not deny their own European perspective, but no longer want to accept it as the only valid "standard". It is an attempt to use words to subject both the tropes of discourse and the discourses of tropes[115] to a self-critique aimed at a multi-logical understanding that is concerned with the most diverse forms and norms of life and coexistence. Adelbert von Chamisso's work on language is work on the myth of *one* logic, one single standard that could be applied to all cultures. Chamisso, however, refuses to apply such a binding standard and to impose it ruthlessly on a world scale on all cultures.

Here, Adelbert von Chamisso does not construct an alterity, an 'other' or 'stranger' that could immediately be inferiorized, but rather attempts to unfold a knowledge of life that is more situated in an epistemology of extension. The experience of the world that manifests itself here is one that aims to live on: That does not want to engage with the narrow notions of 'mannered' or 'civilized' life, but intends a fundamental widening not only of imaginable or conceivable, but of perceptible and interpretable forms of life.

What is at stake here is an experience of the world, such as the Chamisso of 1836 repeatedly projects onto the Chamisso of 1815, thus keeping the game of narrative and experiential I going from a distance of almost twenty years. In his *A Voyage Round the World*, the well-travelled Prussian Adelbert von Chamisso clearly pleads for a further, broader understanding of (human) life.

Again and again, the traveller tries to translate the different logics of Europeans and indigenous people into one another, for example by running towards the people of one of the island groups he visits, waving his hands in the air and shouting, so that they take flight before he succeeds, with the help of his laughter, in attracting the inhabitants again and entering into communication with them.[116] These are

---

Begegnung. Munich: dtv 1982.

[114] Chamisso, Adelbert von: *Reise um die Welt*, vol. II, p. 75.

[115] Cf. on the forms of 'wildness' White, Hayden: *Tropics of Discourse. Essays in Cultural Criticism.* Baltimore/London: the Johns Hopkins University Press 1978, p. 80 f.

[116] Cf. ibid., vol. II, p. 134 f.

(sometimes naive, sometimes ingenious) experiments in living together, which repeatedly test the chances and limits of interpersonal conviviality. What is impressive here is Chamisso's 'gift' or perhaps rather ability to engage with those people whom he knows to understand not as 'savages', not as 'strangers', not as 'others', but as his fellow human beings.

Thus, in the context of his description of the island of Radack, he also referred to his "friend Kadu", who "foreign to this chain of islands" joined the voyage of the 'Rurik' for a time, as one of "the most beautiful characters I have encountered in life", indeed as one of "the people I loved most".[117] The South Sea Islander, as a fellow-man, has become the beloved friend of the French Prussian.

Adelbert von Chamisso remained very true to himself in the construction of his inquisitive explorer ego, the well-travelled world traveller, the multilingual Russian who is not Russian but German, a German who is not German but French, a Frenchman who was not French but a child of Champagne, which he can no longer be, in short: in the construction of an ego which, in its inner-European brokenness, is opened up by his world journey to a further experience of the world. For from the beginning of his world journey he was to fare quite "as at all in the world, where only life teaches life".[118]

This knowledge of life, however, as the result of an open-ended life process and a life story, which—as Humboldt rightly recognized—is far more than a more or less successful description of a journey, opens onto another kind of knowledge that is almost on the threshold of a further life, indeed a continuing life. Thus we read a few pages before the conclusion of the 'Diary' and thus of the first part of Chamisso's *A Voyage Round the World*:

> For my part, with every new chapter of my life, which I end badly and rightly, as well as I can, I am modestly prepared that it will only bring me the wisdom at the end, which I needed at the beginning; and that I will find the missed wisdom of my life on my deathbed.[119]

At the end of the journey, so the hope, after the 'passing away' of all chapters, knowledge could translate into wisdom, knowledge could merge into wisdom and thus live on without being able to be broken down to a concrete knowledge of a science. Only on the pillow of death, according to Chamisso's image, does that wisdom slowly become recognizable that perspectivizes knowledge towards a future, towards a further, towards a continued life.

Wisdom, the aging poet seems to imply at the end of his 'Diary', is at best that knowledge which, at the end of the journey, can only be transformed by death into wisdom—into a wisdom which, after all, would have been so badly needed much earlier, in earlier chapters. Does this mean that this wisdom, as the Franco-Prussian poet seems to imply, cannot be carried on, and thus *lived on*, either transindividually or transgenerationally? In other words: What remains after and of the many journeys of a long life? Can a life that has come to an end not be lived on with all its knowledge and with all its wisdom?

---

[117] Ibid., vol. II, p. 141.

[118] Ibid., p. 20.

[119] Ibid., vol. II, p. 251.

The specific knowledge on the way to wisdom is one that does not seem to get along without death and thus without transcendence. If knowledge, always in the case of the inquisitive, among whom—as we have seen—the author also counts himself, ultimately only results from movement, just as science, which is human in every respect, can only develop *further* from movement, wisdom seems to open itself towards a further life that is located beyond space, beyond time, beyond movement.

But perhaps, we could say to the poet of German Romanticism and the visitor to Rahel Levin's salon, there are *further* paths of knowledge here too that could lead us from science to knowledge and from knowledge further to wisdom. Experiencing the world (Welterleben) and living on (Weiterleben) also seem to meet in that sketch that Roland Barthes placed at the end of his inaugural lecture at the *Collège de France*:

> There is an age when one teaches what one knows; but then comes another age when one teaches what one does not know: it is called seeking. Perhaps now comes the age of another experience: that of disentangling, of letting work the unpredictable reworking that forgetting imposes on the sedimentation of the knowledge, the cultures, the beliefs that one has passed through. This experience has, I believe, an illustrious and unfash-ionable name, which I will give it without being complex, even taking its etymology into account: Sapientia: no power, a little knowledge, a little wisdom, and as much flavor as possible.[120]

The tireless traversal of the most diverse forms of knowledge, cultures and beliefs leads, if we follow Roland Barthes, to a knowledge that, beyond all power, understands wisdom as spice, as a sensual and sensuous experience of the world. Such an experience of the world, so the hope and task of Philology, can be passed on (and not only through editions). Whether it can perhaps also be lived on in all its *saveur* is a question that would have to be posed to the medium of living on *par excellence*, to the Literatures of the World, from ever new perspectives and from further chapters.

The more than one hundred years of Mobile Prussia, which turned and moved in these pages, by no means end with the works of the three Prussian academicians who enriched the play of mobile figures in this last chapter of the present volume. Whether with Anton Wilhelm Amo or Heinrich von Kleist, whether with the protagonists of the Berlin debate about the New World or the Humboldt brothers, whether with Adelbert von Chamisso or the numerous figures in the Berlin salons of Rahel Levin Varnhagen: it was always about a Prussia that was thought considerably further and that can still be thought further today. For that Prussia, which was imaginable and conceivable, writable and to some extent livable, has—we hope—come alive again in these *figurae vitae* of his mobile. If it showed us the views of a Prussia beyond the national, which was not allowed to be and no longer is, it nevertheless holds for us in its manifold overlapping and inconclusive movements a knowledge that points prospectively into the future beyond Prussia with *savoir* and *saveur.*

---

[120] Barthes, Roland: *Leçon. Leçon inaugurale de la Chaire de sémiologie littéraire du Collège de France, prononcée le 7 janvier 1977.* Paris: Seuil 1978, p. 45 f.

Selective Bibliography

The chapters of the present volume are further developments of the following texts by the author:

Anton Wilhem Amo: Philosophieren ohne festen Wohnsitz. Eine Philosophie der Aufklärung zwischen Europa und Afrika. Berlin: Kulturverlag Kadmos 2014.

Cornelius de Pauw, Friedrich II. und die Neue Welt - Oder: Der Sinn der Macht, die Macht über den Sinn und die Macht der Sinne. In: *Jahrbuch 2013 des Staatlichen Instituts für Musikforschung Preußischer Kulturbesitz* (Mainz) (2014), pp. 61–95.

Von Rousseau und Diderot zu Pernety und de Pauw: Die Berliner Debatte um die Neue Welt. In: Dill, Hans-Otto (ed.): *Jean-Jacques Rousseau zwischen Aufklärung und Moderne. Akten der Rousseau-Konferenz der Leibniz-Sozietät der Wissenschaften zu Berlin am 13. Dezember 2012 anlässlich seines 300. Geburtstages am 28. Juni 2012 im Rathaus Berlin-Mitte.* Berlin: Leibniz-Sozietät der Wissenschaften (= Sitzungsberichte der Leibniz-Sozietät der Wissenschaften 117) 2013, pp. 111–130.

Alexander von Humboldt und die Globalisierung: Das Mobile des Wissens. Frankfurt a. M.: Insel 2009, pp. 97–127.

Kleist – Karibik – Konvivenz. 'Die Verlobung in St. Domingo' als Erprobungsraum künftigen Zusammenlebens. In: Blänkner, Reinhard (Hg.): *Heinrich von Kleists Novelle, Die Verlobung in St. Domingo. Literatur und Politik im globalen Kontext um 1800.* Würzburg: Königshausen & Neumann 2013, pp. 187–224.

Festvortrag: "Sprachen über Sprachen. Zwei Brüder und eine Humboldt'sche Wissenschaft." In: *Jahrbuch 2017. Berlin: Berlin-Brandenburgische Akademie der Wissenschaften* 2018, pp. 202–218.

Welterleben/Weiterleben. Zur Vektopie bei Georg Forster, Alexander von Humboldt und Adelbert von Chamisso. In: Drews, Julian/Ette, Ottmar/Kraft, Tobias/Schneider-Kempf, Barbara/Weber, Jutta (ed.): *Forster – Humboldt – Chamisso. Weltreisende im Spannungsfeld der Kulturen.* Mit 44 Abbildungen. Göttingen: V&R unipress 2017, pp. 383–427.

Amo Afer aus Axim in Ghana, Antonius Guilielmus: *Übersetzung seiner Werke.* [Übers. A. Blaschka?]. Halle (Saale): Martin-Luther-Universität Halle-Wittenberg 1965.

Amo, Anton Wilhelm: *Tractatus de arte sobrie et accurate philosophandi.* Halle: Kitler 1738.

Arendt, Hannah: *Rahel Varnhagen. Lebensgeschichte einer deutschen Jüdin aus der Romantik.* München/Zürich: Piper 1981.

Assing, Ludmilla (ed.): *Briefe von Alexander von Humboldt an Varnhagen von Ense aus den Jahren 1827 bis 1858. Nebst Auszügen aus Varnhagen's Tagebüchern und Briefen von Varnhagen und Andern an Humboldt.* Leipzig: F. A. Brockhaus 1860.

Beck, Hanno: *Alexander von Humboldt.* 2 Bde. Bd. 1. Wiesbaden: Franz Steiner 1959.

O. Ette, *Mobile Prussia*, https://doi.org/10.1007/978-3-476-05839-3

Bergvelt, Elinoor/Wezel, Elsa van (eds.): *Museale Spezialisierung und Nationalisierung ab 1830. Das Neue Museum in Berlin im internationalen Kontext./Spezialisation and Consolidation of the National Museum after 1830.* The Neue Museum in Berlin in an International Context. Berlin: G + H 2011.

Bernaschina, Vicente/Kraft, Tobias/Kraume, Anne (eds.): *Globalisierung in Zeiten der Aufklärung. Texte und Kontexte zur ‚Berliner Debatte' um die Neue Welt (17./18. Jh.).* 2 Teile. Frankfurt a. M./Bern/New York: Peter Lang Edition 2015.

Blamberger, Günter: *Heinrich von Kleist.* Biographie. Frankfurt a. M.: S. Fischer 2011.

Bourguet, Marie-Noëlle: *Le monde dans un carnet. Alexander von Humboldt en Italie (1805).* Paris: Editions du Félin 2017.

Brentjes, Burchard: *Anton Wilhelm Amo. Der schwarze Philosoph in Halle.* Leipzig: Koehler & Amelang 1976, p. 32/33.

Broc, Numa *: La Géographie des Philosophes. Géographes et voyageurs français au XVIIIe siècle.* Paris: Editions Ophrys 1975.

Chamisso, Adelbert von: *Leben und Briefe.* Hg. von Julius Eduard Hitzig. Leipzig: Weidmann'sche Buchhandlung 1942.

Chamisso, Adelbert von: *Sämtliche Werke.* München: Winkler 1975.

Danzer, Gerhard: *Europa, deine Frauen. Beiträge zu einer weiblichen Kulturgeschichte.* Berlin/ New York: Springer 2015.

Detering, Susanne: *Kolumbus, Cortés, Montezuma: Die Entdeckung und Eroberung Lateinamerikas als literarisches Sujet in der Aufklärung und im 20.* Jahrhundert. Weimar: Verlag und Datenbanken für Geisteswissenschaften 1996.

Ette, Ottmar: *Alexander von Humboldt Handbuch.* Stuttgart: J.B. Metzler 2018a.

Ette, Ottmar (ed.): *Das Buch der Begegnungen: Menschen – Kulturen – Geschichten aus den Amerikanischen Reisetagebüchern.* München: Manesse 2018b.

Ette, Ottmar: *WeltFraktale. Wege durch die Literaturen der Welt.* Stuttgart: J. B. Metzler 2017.

Fischer, Robert: *Adelbert von Chamisso. Weltbürger, Naturforscher und Dichter. Vorwort von Rafik Schami.* Mit zahlreichen Abbildungen. Berlin/München: Erika Klopp 1990.

Floeck, Wilfried/Fritz, Sabine (eds.): *La representación de la Conquista en el teatro español desde la Ilustración hasta finales del franquismo.* Hildesheim/Zürich/New York: Georg Olms 2009.

Fontius, Martin: *Friedrich II. und die europäische Aufklärung.* Berlin: Duncker & Humblot 1999.

Forster, Georg: *Werke. Sämtliche Schriften, Tagebücher, Briefe.* Hg. von Horst Fiedler, Klaus-Georg Popp, Annerose Schneider und Christian Suckow. Berlin: Akademie-Verlag 1985a.

Forster, Georg: *Ansichten vom Niederrhein.* München: The Hilliard Collection – Edition Deutsche Bibliothek 1985b.

Forster, Georg: *Reise um die Welt.* Edited and commented with an afterword by Gerhard Steiner. Frankfurt a. M.: Insel 1983.

Friedrich II.: *Anti–Machiavel ou Examen du Prince de Machiavel.* Corrigé pour la plus grande partie d'après le manuscrit original de Frédéric II. Avec une introduction et des notes historiques. Hambourg: Chez Frédéric Perthes 1834.

Friedrich II.: *Correspondance de Frédéric Second, Roi de Prusse, avec le Comte Algarotti.* Pour servir de suite aux éditions des œuvres posthumes de ce Prince. Berlin: Chez George Gropius, Libraire 1887.

Friedrich II./Tagliazucchi, Giuseppe: *Montezuma. Tragedia per musica.* Da rappresentarsi nel Regio Teatro di Berlino per Ordine di Sua Maestà il Re nel Carnovale del MDCCLV. Berlino: Appresso Haude e Spener con licenza di Sua Maestà 1755.

Gerbi, Antonello: *La Disputa del Nuovo Mondo. Storia di una Polemica: 1750–1900.* Nuova edizione a cura di Sandro Gerbi. Milano/Napoli: Riccardo Ricciardi Editore 1983.

Gier, Albert (ed.): *Oper als Text: Romanistische Beiträge zur Libretto-Forschung.* Heidelberg: Winter 1986.

Gordon, Lewis R.: *An Introduction to African Philosophy.* Cambridge: Cambridge University Press 2008.

Grunewald, Michel/Lüsebrink, Hans-Jürgen/Marcowitz, Reiner/Puschner, Uwe (eds.): *France-Allemagne au XXe siècle. La production de savoir sur l'Autre.* Bern/Frankfurt a. M./New York: Peter Lang 2012.

Günther, Horst (ed.): *Die Französische Revolution. Berichte und Deutungen deutscher Schriftsteller und Historiker.* Frankfurt a. M.: Deutscher Klassiker Verlag 1988.

Hahn, Barbara: *'Im Schlaf bin ich wacher.' Die Träume der Rahel Levin Varnhagen.* Frankfurt a. M.: Luchterhand 1990.

Hamel, Jürgen, Knobloch, Eberhard/Pieper, Herbert (eds.): *Alexander von Humboldt in Berlin. Sein Einfluß auf die Entwicklung der Wissenschaften.* Beiträge zu einem Symposium. Augsburg: ERV 2003.

Henze-Döhring, Sabine: *Friedrich der Große: Musiker und Monarch.* München: C.H. Beck 2012.

Humboldt, Alexander von: *Aus meinem Leben. Autobiographische Bekenntnisse.* Selected and commented by Kurt-R. Biermann. München: C.H. Beck 1987.

Humboldt, Alexander von: *Briefe aus Amerika 1799–1804.* Bearbeitet von Ulrike Moheit. Berlin: Akademie-Verlag 1993.

Humboldt, Alexander von: *Reise auf dem Río Magdalena, durch die Anden und Mexiko. Teil I: Texte.* Hg. von Margot Faak. With an introductory study by Kurt R. Biermann. Zweite, durchgesehene und verbesserte Auflage. Berlin: Akademie-Verlag 2003.

Humboldt, Alexander von: *Relation historique du Voyage aux Régions équinoxiales du Nouveau Continent [...].* Reprint of the complete original published in Paris in 1814–1825, procured, introduced and augmented with an index by Hanno Beck. Stuttgart: Brockhaus 1970.

Humboldt, Alexander von: *Vues des Cordillères et Monumens des Peuples Indigènes de l'Amérique.* Nanterre: Editions Erasme 1989.

Humboldt, Wilhelm von: *Briefe.* Edited and commented by Phillip Mattson. Berlin: de Gruyter 2015.

Kasten, Erich: Reisen an den Rand des Russischen Reiches: *Die wissenschaftliche Erschließung der nordpazifischen Küstengebiete im 18. und 19. Jahrhundert.* Fürstenberg/Havel: Kulturstiftung Sibirien 2013.

Kleist, Heinrich von: *Sämtliche Werke und Briefe in vier Bänden.* Edited by Ilse-Marie Barth, Klaus Müller-Salget, Stefan Ormanus and Hinrich C. Seeba. Frankfurt a. M.: Deutscher Klassiker Verlag 1987.

Kunisch, Johannes: *Friedrich der Große. Der König und seine Zeit.* With 29 images and 16 maps. München: Deutscher Taschenbuch Verlag^2 2010.

Leithold, Norbert: *Friedrich II. von Preußen. Ein kulturgeschichtliches und bebildertes Panorama von A bis Z.* Frankfurt a. M.: Eichborn (Die Andere Bibliothek) 2011.

Lietzmann, Albert (ed.): *Wilhelm von Humboldts Gesammelte Schriften. Tagebücher.* Berlin: de Gruyter 1968.

Lüsebrink, Hans-Jürgen (ed.): *Das Europa der Aufklärung und die außereuropäische koloniale Welt.* Göttingen: Wallstein 2006.

Lüsebrink, Hans-Jürgen/Strugnell, Anthony (eds.): *L'‚Histoire des deux Indes': réécriture et polygraphie.* Oxford: Voltaire Foundation 1996.

Mabe, Jacob Emmanuel: *Anton Wilhelm Amo interkulturell gelesen.* Nordhausen: Traugott Bautz 2007.

Pauw, Cornelius de: *Recherches philosophiques sur les Américains, ou Mémoires intéressants pour servir à l'Histoire de l'Espèce humaine.* 2 Bde. Berlin: Georges Jacques Decker, Imp. du Roi 1768–1769.

Pernety, Antoine-Joseph: *Examen des Recherches philosophiques sur les Américains et de la Défense de cet ouvrage.* 2 Bde. Berlin: Georges Jacques Decker, Imprimeur du Roi 1771.

Preuss, Johann D. (ed.): *Oeuvres de Frédéric le Grand.* Tome XXI. Berlin: Imprimerie Royale (R. Decker) 1853.

Raynal, Guillaume-Thomas: *Histoire philosophique et politique des établissements et du commerce des européens dans les deux Indes.* Genève: Chez Jean-Léonard Pellet, Imprimeur de la Ville & de l'Académie 1781.

Rebok, Sandra: *Alexander von Humboldt und Spanien im 19. Jahrhundert: Analyse eines wechselseitigen Wahrnehmungsprozesses.* Frankfurt a. M.: Vervuert 2006.

Sproll, Monika: *Adelbert von Chamisso in Cunersdorf.* Frankfurt (Oder): Kleist-Museum 2014.

Stern, Carola: *Der Text meines Herzens. Das Leben der Rahel Varnhagen.* Reinbek bei Hamburg: Rowohlt 1994.

Uerlings, Herbert: *Poetiken der Interkulturalität. Haiti bei Kleist, Seghers, Müller, Buch und Fichte.* Tübingen: Max Niemeyer 1997.

Varnhagen, Rahel: *Briefe und Aufzeichnungen.* Hg. von Dieter Bähtz. Frankfurt a. M.: Insel 1986.

Varnhagen, Rahel: *Gesammelte Werke.* München: Matthes & Seitz 1983.

Volz, Gustav Berthold (ed.): *Friedrich der Große: Historische, militärische und philosophische Schriften, Gedichte und Briefe.* Mit Illustrationen von Adolph von Menzel. Hg. von Gustav Berthold Volz. Köln: Anaconda 2006.

Zantop, Susanne: *Colonial Fantasies. Conquest, Family, and Nation in Precolonial Germany, 1770–1870.* Durham/London: Duke University Press 1997.

Zeuske, Michael: *Handbuch Geschichte der Sklaverei. Eine Globalgeschichte von den Anfängen bis zur Gegenwart.* Berlin/Boston: de Gruyter 2013.

FSC
www.fsc.org
MIX
Papier | Fördert
gute Waldnutzung
FSC® C083411

Kolibri 360 GmbH
Ferdinand-Jühlke-Straße 7
99095 Erfurt, Deutschland
produktsicherheit@kolibri360.de